SECOND EDITION

Engineering Design:
A Project-Based Introduction

CLIVE L. DYM
PATRICK LITTLE

Harvey Mudd College

JOHN WILEY & SONS, INC.

To

Joan Dym
whose love and support are distinctly nonquantifiable

and to

Charlie Hatch
a teacher's teacher

ACQUISITIONS EDITOR   Joseph Hayton
ASSOCIATE MARKETING DIRECTOR   Ilse Wolfe
SENIOR PRODUCTION EDITOR   Ken Santor
SENIOR DESIGNER   Kevin Murphy
ILLUSTRATION COORDINATOR   Sandra Rigby
COVER DESIGNER   Miriam Dym

This book was set in Times Roman by Argosy Publishing Inc. and printed and bound by Malloy, Inc. The cover was printed by Phoenix Color.

This book is printed on acid-free paper.

To order books please call 1(800)-225-5945.

ISBN 0-471-25687-0

Printed in the United States of America

10 9 8 7 6

FOREWORD

To design is to imagine and specify things that don't exist, usually with the aim of bringing them into the world. The "things" may be tangible—machines and buildings and bridges; they may be procedures—the plans for a marketing scheme or an organization or a manufacturing process, or for solving a scientific research problem by experiment; they may be works of art—paintings or music or sculpture. Virtually every professional activity has a large component of design, although usually combined with the tasks of bringing the designed things into the real world.

Design has been regarded as an art, rather than a science. A science proceeds by laws, which can sometimes even be written in mathematical form. It tells you how things must be, what constraints they must satisfy. An art proceeds by heuristic, rules of thumb, and "intuition" to search for new things that meet certain goals, and at the same time meet the constraints of reality, the laws of the relevant underlying sciences. No gravity shields; no perpetual motion machines.

For many years after World War II, science was steadily replacing design in the engineering college curricula, for we knew how to teach science in an academically respectable, that is, rigorous and formal, way. We did not think we knew how to teach an art. Consequently, the drawing board disappeared from the engineering laboratory—if, indeed, a laboratory remained. Now we have the beginnings—more than the beginnings, a solid core—of a science of design.

One of the great gifts of the modern computer has been to illuminate for us the nature of design, to strip away the mystery from heuristics and intuition. The computer is a machine that is capable of doing design work, but in order to learn how to use it for design, an undertaking still under way, we have to understand what the design process is.

We know a good deal, in a quite systematic way, about the rules of thumb that enable very selective searches through enormous spaces. We know that "intuition" is our old friend "recognition," enabled by training and experience through which we acquire a great collection of familiar patterns that can be recognized when they appear in our problem situations. Once recognized, these patterns lead us to the knowledge stored in our memories. With this understanding of the design process in hand, we have been able to reintroduce design into the curriculum in a way that satisfies our need for rigor, for understanding what we are doing and why.

One of the authors of this book is among the leaders in creating this science of design and showing both how it can be taught to students of engineering and how it can be implemented in computers that can share with human designers the tasks of carrying out the design process. The other is leading the charge to integrate the management sciences into both engineering education and the successful conduct of engineering design projects. This book thus represents a marriage of the sciences of design and of management. The science of design continues to move rapidly forward, deepening our understanding and enlarging our opportunities for human-machine collaboration. The study of design has joined the study of the other sciences as one of the exciting intellectual adventures of the present and coming decades.

Herbert A. Simon
Carnegie Mellon University
August 6, 1998

PREFACE

Why should you read this book on engineering design?

In writing both editions of this book we confronted many of the same issues that we discuss in the pages that follow. It was important for us to be very clear about our overall objectives, which we outline below, and about the particular objectives we had for each chapter. We asked about the pedagogic function served by the various examples, and whether some other example or tool might provide a better means for achieving that pedagogical function. The resulting organization and writing represents the implementation of our best design. Thus, this and all books are definitely designed artifacts: They require the same concern with objectives, choices, functions, means, budget, and schedule, as do other engineering or design projects.

There are a lot of books on design, engineering design, project management, team dynamics, project-based learning, and the other topics we cover in this volume (see the Bibliography). This book grew out of our desire to combine these topics in a single, introductory work that focused particularly on conceptual design. That desire arose from our teaching at Harvey Mudd College, where our students do team-based design projects in a first-year design course, "Introduction to Engineering Design," that we affectionately call "E4," and in the Engineering Clinic. Clinic is an unusual "capstone course" taken during junior (for one semester) and senior years (for both semesters) in which students work on externally-sponsored design and development projects. In both E4 and Clinic, Mudd students work in multi-disciplinary teams, under specified time deadlines, and within specified budget constraints. These conditions replicate to a significant degree the environments within which most practicing engineers will do most of their professional design work. In looking for books that could serve our audience, we found that there were excellent texts covering detailed design, usually targeted toward senior capstone design courses, or "introductions to engineering" that focused on detailing the branches of engineering. We could not find a book that introduced the processes and tools of conceptual design in a project or team setting that did not presume substantial engineering knowledge.

This book is directed to three related audiences: students, teachers, and practitioners. While each group has its own special concerns, those interests are intimately tied together by the topics of this book.

The book is intended to enable *students* to learn the nature of design, the central activity of engineering, either directly or as an ultimate aim. (Even the most focused materials engineers, for example, hope that any new materials they develop will be used in the design of *something*.) We also hope the book will help students learn the tools and techniques of formal design that will be useful in framing the design problems they will face during their education and during their careers. the issues and tools of project management will also be faced by graduates when they go on to careers in industry, government, and academia. Since design (and research) are increasingly done in teams, the insights and tips on team dynamics will be valuable as young engineers reflect on their occupational life as well. We have included examples of work done by our students on actual projects in E4, both to show how the tools are used and to highlight some frequently made mistakes. We hope student readers will take heart from these illustrations and use them in learning to be effective engineers (or at least effective engineering students).

Given that we are professors at an undergraduate college of engineering and science, we wrote this book with *teachers* very much in mind. We considered both the delivery of this material to students and the ways in which professors might teach introductory design courses. Thus, this book is structured to allow a teacher to use ongoing examples for illustration and as homework (or in-class) exercises. The material is ordered so that professors can decide for themselves whether to cover the ideas in the text prior to or concurrently with particular stages in design projects. We have tried both approaches in our courses and find that each has its benefits. An accompanying *Instructor's Manual* outlines sample syllabi and organizations for teaching the material in the book, as well as additional examples. We also discuss the structure of the material in the book later in the preface to help teachers decide how to use the book.

Finally, we hope the book will be useful to *practitioners*, either as a refresher of things learned or as an introduction to some essential elements of conceptual design that were not formally introduced in engineering curricula in years past. We do not assume that the examples given here substitute for an engineer's experience, but do believe that case studies show the relevance of these tools to practical engineering settings. Some of our friends and colleagues in the profession like to point out that the tools we teach would be unnecessary if only we all had more common sense. Notwithstanding that, the number and scale of failed projects suggests that common sense may not, after all, be so commonly distributed. In any case, this book offers both practicing engineers (and engineering managers) a view of the design tools that even the greenest of engineers will have in their toolbox in the coming years.

SOME SPECIFICS ABOUT WHAT'S COVERED

Design is an *open-ended* and ill-structured process, by which we mean there is no unique solution, and that the candidate solutions cannot be generated with an algorithm. As we discuss in the early chapters, designers have to provide an orderly process for organizing an ill-structured design activity in order to support making decisions and tradeoffs among competing possible solutions. As such, algorithms and mathematical formulations cannot replace the imperative to understand the needs of various stakeholders (clients, users, and the public), even if those mathematical tools are used later in the design process. This lack of structure and available formal mathematical tools make the introduction of conceptual design early in the curriculum possible and, we think, desirable. It provides a framework in which engineering science and analysis can be used, while not demanding skills that most first- and second-year students have not yet acquired. We have, therefore, included in this book the following specific tools for conceptual design, for acquiring and organizing design knowledge, and for managing the team environment in which design takes place.

The following *formal conceptual design methods* are delineated:

- objectives trees
- pairwise comparison charts
- functional analysis
- function-means trees
- morphological charts
- requirements matrices
- performance specifications
- quality function deployment (QFD)

Since conceptual design thinking requires and produces a lot of information, we also discuss several *means of acquiring and processing information*:

- literature reviews
- brainstorming
- synectics and analogies
- user surveys and questionnaires
- benchmarking
- reverse engineering (or, dissection)
- metric definitions
- laboratory experiments
- simulation and computer analysis
- formal design reviews

Since the successful completion of any design project by a team requires that team members estimate a project's scope of work, schedule, and resources early in the life of the project, we also introduce several *design management tools*:

- work breakdown structures
- linear responsibility charts
- schedules
- activity networks
- Gantt charts
- budgets
- control tools

INTEGRATIVE DESIGN EXAMPLES

As a rare feature, we use one case study and two integrative examples to follow the design process through to completion, thus showing each of the tools and techniques as they are used on the same design project. In addition to numerous "one-time" examples, we detail the following case study and integrative examples:

1. Design of a ***microlaryngeal surgical stabilizer***, a device used to stabilize the instruments during throat surgery. This case study derives from a Harvey Mudd College first-year design project sponsored by the Beckman Laser Institute of the University of California Irvine.

2. Design of a ***beverage container***. The designer, having a fruit juice company as a client, is asked to develop a means of delivering a new beverage to a market predominantly composed of children and their parents. There are clearly a number of possibilities (e.g., mylar bags, molded plastics), and issues such as environmental effects, safety, and the costs of manufacturing are considered.

3. Design of a chicken coop to be built and used by a Mayan cooperative in Guatemala. The chicken coop was designed by a team of Harvey Mudd College students in our E4 design course. It was subsequently built by the students, on-site and of indigenous materials, with support from a humanitarian aid group called Xela-Aid.

In addition, the accompanying *Instructor's Manual* includes a case study of the design of a ***transportation network*** to enable automobile commuter traffic between Boston and its northern suburbs, through Charlestown, Massachusetts. This conceptual design problem clearly illustrates the many factors that go into large-scale engineering projects in their early stages, when choices are being made between highways, tunnels, and bridges. Among the design concerns are cost, implications for future expansion, and preservation of the character, environment, and even the view of the affected neighborhoods. This project is also an example of how conceptual design thinking can significantly influence some very "real world" events.

We also cover several topics that may be viewed as less central in a first exposure to design, but we feel they are very important. Thus, in Chapter 6 we discuss the endgame and completion of a design project with a strong emphasis on the ways and means of reporting design results. In Chapter 8 we discuss "design for X" issues, including manufacturing and assembly, affordability (engineering economics), reliability and maintainability, sustainability, and quality. Finally, we discuss various ethics issues in design in Chapter 9.

Clive L. Dym
Patrick Little
January 2003

ACKNOWLEDGMENTS

A book like this does not get written without the faith, support, advice, criticism, and help from many people. We want to thank some of those people, as follows:

Our colleagues and our students at Harvey Mudd College, especially those who participated in our E4 course since it was introduced in its current form in the Spring 1992. *Rich Phillips*, recent department chair (1993 to 1999), for working to make E4 a signature course for the engineering department by teaching it himself several times and by ensuring that most of our colleagues served on an E4 teaching team at least once. *Jim Rosenberg*, for exceptional generosity in sharing his experiences and insights from having taught the course several times. *Joe King* for reviewing our overview of the role of drawings in design and to the HMC alums Michael James Messina and Philip Johnson for rendering some of the figures. We are also grateful to the following student teams for allowing us to use their design results:

- *Thomas Both, Genevieve Breed, Chris Stratton,* and *Kristen Van Horn* (Both et al. 2000);
- *Stephanie Chan, Ryan Ellis, Micah Hanada,* and *Judy Hsu* (Chan et al. 2000);
- *Jeannie Connor, Kristina Kubler, Peter Leitzell, J. P. Strozzo,* and *Mark Wang* (Connor et al. 1997);
- *Lance Feagan, Tom Galvani, Shannon Kelley,* and *Markus Ong* (Feagan et al. 2000);
- *Peter Gutierrez, Joey Kimball, Brian Maul, Adam Thurston,* and *Jake Walker* (Gutierrez et al. 1997); and
- *Yanos Saravanos, Justin Schauer,* and *Clifford Wassman* (Saravanos et al. 2000).

Ted Belytschko, Ed Colgate, Leon Keer, and *Greg Olson* of Northwestern University for inviting CLD to serve as Eshbach Visiting Professor of Civil Engineering at Northwestern during 1997 to 1998, where much of the first edition of this manuscript began. Ted and Leon handled the administrative stuff, Ed provided great support while leading the development of a new first-year course ("Engineering Design and Communication,"), and Greg was, and still is, a stimulating interlocutor on all sorts of design issues.

Miriam Dym for designing the covers for both editions.

Peter R. Frise (University of Windsor), *Larry G. Richards* (University of Virginia), *Susan Carlson Skalak* (University of Virginia) and *Burt L. Swersey* (Rensselaer Polytechnic Institute) for their suggestions and insights. Particular thanks to Peter, Larry, and Susan for very constructive comments that helped shape our thinking during the writing of the first edition.

Joseph Hayton of John Wiley & Sons for his continuing support of this project from its inception.

J. Stanley Johnson and *Mary Wig Johnson* for providing Harvey Mudd with the endowed chair from which PL was able to write this book, and, even more significantly, for tasking PL to look for ways to incorporate management into engineering education.

William J. LeMessurier of William J. LeMessurier Consultants for reviewing Chapter 9 and providing some very helpful insights.

Herbert A. Simon of Carnegie Mellon University for providing one of us (CLD) with much encouragement over many years and for generously penning the foreword to this book.

Finally, *Joan Dym* and *Judy Little*, our wives, for tolerating us during the absences that such projects entail, and for listening to each of us as we worked through differences to find a common voice.

CONTENTS

PHOTO CREDITS

Chapter 1
Fig. 1.2a: Courtesy Lockheed Advanced Development Company
Fig. 1.2b: Courtesy British Airways
Fig. 1.2c: Courtesy The Boeing Corporation
Fig. 1.2d: Courtesy AeroVironment, Inc.
Fig. 1.3a: John Welzenbach/Corbis Stock Market
Fig. 1.3b: Jose Carrillo/ PhotoEdit/ Picturequest
Fig. 1.3c: Phyllis Picardi/Stock Boston/ Picturequest
Fig. 1.3d: Alan Levenson/Stone/Getty Images
Fig. 1.5a: Terje Rakke/The Image Bank/Getty Images
Fig. 1.5b: Pascal Crapet/Stone/Getty Images
Fig. 1.5c: Yeager/Corbis Stock Market
Fig. 1.5d: G.K. & Vikki Hart/The Image Bank/Getty Images

Chapter 5
Fig. 5.1: Courtesy Johnson & Johnson
Fig. 5.10: Courtesy Clive L. Dym

Chapter 6
Fig. 6.6: Courtesy Clive L. Dym
Fig. 6.7: Courtesy Clive L. Dym

Chapter 9
Fig. 9.3: Courtesy Clive L. Dym
Fig. 9.4: Courtesy Clive L. Dym

ENGINEERING DESIGN

*What does it mean to design something? How does
engineering design differ from other kinds of design?*

PEOPLE HAVE been designing things for as long as we can "remember" or archaeologically uncover. Our earliest ancestors designed flint knives and other basic tools to help meet their most basic needs. They also designed wall paintings to tell stories and make their primitive caves visually more comfortable. Given the long history of people designing *artifacts*, it is useful to ask how an engineer designing structural members to support a building's occupants is somehow different from the interior designer of carpets and wall hangings to decorate that building. We will use this chapter to set some contexts for engineering design and to begin developing both a vocabulary and a shared understanding of what we mean by engineering design.

1.1 WHERE AND WHEN DO ENGINEERS DESIGN?

What does it mean for an *engineer* to design something? When do engineers design things? Where? Why? For whom?

There are a lot of questions we could ask about engineers doing design, and there are probably more answers than there are questions. An engineer could be working for a large company that processes and distributes various food products, in which case she might be asked to design a new container for a new juice product. He could be working for a design and construction company designing some part of a bridge for a new interstate highway that is part of a larger transportation project. Or, an engineer might be working for an automobile company that wants to develop a new concept for the instrumentation cluster in its cars, perhaps one that makes it easier for drivers to check various parameters without having to take their eyes off the road. Or, the engineer might be working for a school system that wants specialized facilities to better serve students with various kinds of orthopedic disabilities.

Clearly, this is a list that could easily be lengthened, which makes it worth asking whether there are any common elements in the engineers' situations or in the ways that they approach their tasks. In fact, there are common features in both situations and tasks, and it is the existence of such commonalities that makes it possible to describe the design process and the context in which it occurs.

First, we can identify three "roles" being played as the design of a product unfolds. Obviously, there is the *designer*. It seems equally clear that there will be a *client*, the person or group or company that wants a design conceived. For the working engineer, the client could be internal (e.g., the person who decides the food company should start selling a new juice product) or external (e.g., the government agency that contracts for the new highway system). And while the designer may relate differently to internal and external clients, in either case it is the client who presents a project statement from which all else begins to flow. Design project statements are often verbal, and sometimes they are quite short. These two qualities suggest that the designer's first task is to clarify what the client really wants and translate it into a form that is useful to her as an engineering designer. We will discuss more about this in Chapter 3 and beyond, but we want to recognize for now that *a design is motivated by a client* who wants some sort of artifact.

Design is motivated by a client.

There's another player or stakeholder in the design effort and that's the *user*, the person (or the set of people) who will actually use the device or artifact being designed. In the contexts mentioned above, the users are the consumers who buy the new juice, drivers on the interstate highway system, drivers of the new line of autos, and orthopoedically disabled students (and their teachers). The users hold a stake in the design process because a product *won't sell* if its design doesn't meet their needs. Thus, the designer, the client, and the user form a triangle, as shown in Figure 1.1. The designer needs to understand what the client wants, but the client also has to understand what his users need and communicate that to the designer. In Chapter 3 we will describe design processes that model how the designer can interact and communicate with both the client and potential users to help inform her own design thinking. We will also identify tools (discussed in Chapters 4–7) that she can use to organize and refine her thinking.

The designer and the client have to understand what the users want.

Engineering designers work in many different kinds of environments. It is hard to be specific about a designer's working environment because it could include companies large and small, such as start-up ventures, government and not-for-profit agencies, or engineering services firms, one breed of which is the industrial design consultancy. Apart from the salaries and perks that come from working in these various places, the designer will most likely also see differences arising due to the size of a project, the number of colleagues on the design team, and the designer's access to relevant information about user needs. On large projects, many of the designers will be working on pieces of a project that are so detailed and confined that much of what we describe in this book may not seem immediately useful. Thus, the designers of a bridge abutment, an airplane's fuel tank, or components on a computer motherboard will likely not be as concerned with the larger picture of what external clients and users want. Indeed, as we will explain in Chapter 3, these kinds

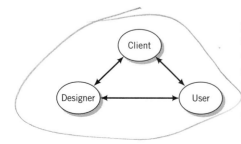

FIGURE 1.1   The designer–client–user triangle. There are three parties involved in a design effort: the client, who has objectives that the designer must clarify; the user of the designed device, who has his own requirements; and the designer, who must develop specifications such that something can be built to satisfy everybody.

of design problems are the part of the process called *detailed design*, in which the choices and procedures are so well understood that more general design issues have already been taken into account. However, even for such large projects, the response to a client's project statement is initiated with *conceptual design*. Some thinking about the size and mission of the airplane has already been done in order to determine the constraints surrounding the design of fuel tanks, while the performance parameters that the computer motherboard must display will be determined by some assessment of the market for and the price of the computer in question.

Large, complex projects often lead to very different interpretations of client project statements and user needs. One has only to look at the many different kinds of skyscrapers that decorate our major cities to see how architects and structural engineers envisage different ways of housing people in offices and apartments. Visible differences also emerge in airplane design (Figure 1.2) and wheelchair design (Figure 1.3). Each of these devices can follow from a simple, common design statement: Airplanes are "devices that safely transport people and goods through the air," and wheelchairs are "personal mobile devices that transport people who are unable to use their legs." However, the different products that have emerged represent different conceptualizations of what the clients and users wanted from these artifacts. Designers thus have much to do when clarifying what a client wants and translating those wants into an engineered product.

The players' triangle also prompts us to (a) consider that the interests of the three players could diverge, and (b) recognize that the consequences of such divergence could mean more than financial problems resulting from a failure to meet users' needs. This is because the interaction of multiple interests creates an interaction of multiple obligations, and these obligations may well conflict. For example, the designer of a juice container might consider metal cans, but easily "squashed" cans are a hazard if sharp edges emerge during the squashing. There might be tradeoffs among design variables, including the material of which a container is to be made and the container's thickness. The choices made in the final design could easily reflect different assessments of the possible safety hazards, which in turn lays the foundation for potential ethics problems. Ethics problems, which we discuss in Chapter 9, occur because the designer has obligations not just to the client and the user, but also to his profession and, as detailed in the codes of ethics of engineering societies, to the public at large. Thus, ethics issues are never terribly far from the design process.

Designers have obligations to the profession and to the public.

Another aspect of engineering design practice that is increasingly common in projects and firms of all sizes is the use of *teams* to do design. Many engineering problems are inherently multidisciplinary (e.g., the design of medical instrumentation), so there is a need to understand the requirements of clients, users, and technologies in very different environments. This, in turn, requires the assembling of teams that can address such different sets of environmental needs. The widespread use of teams clearly affects the management of design projects, another recurring theme of this book.

Engineering design is a multifaceted subject, and in no way do we pretend that we can fully explain this wonderfully complex activity in just one short book. However, we can model ways of thinking productively about some of the conceptual issues and the resulting choices that are made very early in the design of many different kinds of engineered artifacts.

FIGURE 1.2   "Devices that safely transport people and goods through the air," i.e., airplanes. No surprises here, right? We have all seen a lot of airplanes (or, at least we have in pictures or movies). But even these planes, albeit of different eras and origins, clearly show they were designed to achieve very different missions.

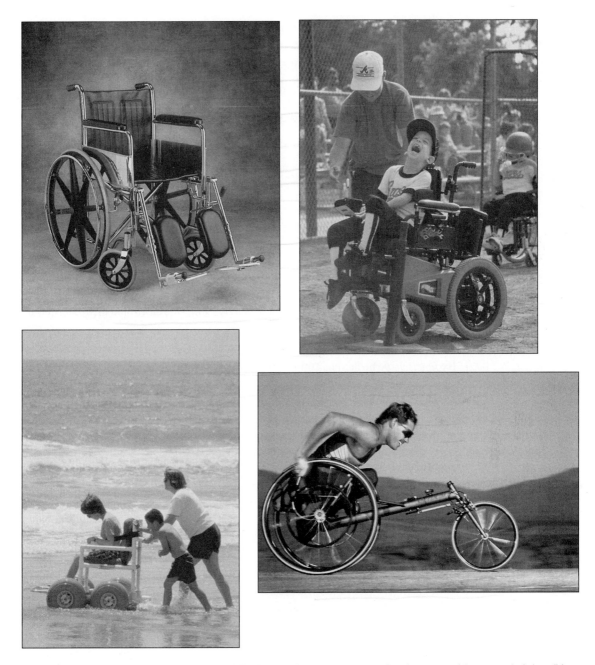

FIGURE 1.3 A collection of "personal mobile devices that transport people who are unable to use their legs," i.e., wheelchairs. Here too, as with the airplane, we see some sharp differences in the configurations and components of these wheelchairs. Why are the wheels so different? Why are the wheelchairs so different?

1.2   DEFINING ENGINEERING DESIGN

Many definitions of engineering design have been offered over the years. We now define and explain what *we* mean by engineering design.

1.2.1   Our definition of engineering design

The following formal definition of engineering design is the most useful one for our purposes:

Engineering design is the systematic, intelligent generation and evaluation of specifications for artifacts whose form and function achieve stated objectives and satisfy specified constraints.

Engineering design is a *thoughtful* process for generating designs that achieve objectives within specified constraints.

What precisely does this mean?

First of all, we define some terms:

- *Artifacts* are human-made objects, the "things" or devices that we are designing. They are most often "physical," like airplanes, wheelchairs, ladders, and carburetors. But "paper" products such as drawings, plans, computer software, articles, and books, are also artifacts. Further, artifacts are increasingly "soft," the electronic files that become "real" when displayed on a computer screen.

- The *form* of an artifact is its shape, its geometry.

- By *function* we mean those things the artifact is supposed to do.

- The *specifications* of artifacts are precise descriptions of the properties of the object being designed. They are typically numerical values of *performance parameters* (i.e., constants or variables that serve as indicators of the artifact's behavior) or of *attributes* (i.e., properties or characteristics of the artifact). For example, we could associate two lengths with an extension ladder: An unextended length of five feet defines a geometric attribute, and the requirement that the ladder enables access to ten foot heights is a performance specification.

- We call the set of values that articulates what a design is intended to do the *design specifications*. They also provide a basis for *evaluating* proposed designs: Design specifications provide the "targets" against which we can measure our success in achieving them.

The definition of design states that the design specifications are found as the result of *systematic, intelligent generation.* This is not to deny that design is a creative process. However, at the same time, there are techniques and tools we can use to support our creativity, to help us think more clearly, and make better decisions along the way. These tools and techniques, which form much of the subject of this book, are not formulas or algorithms. Rather, they are ways of asking questions, and presenting and viewing the answers to those questions as the design process unfolds. We will also present some tools and techniques for managing a design project. Thus, while demonstrating ways of thinking about a design as it unfolds in our heads, we also demonstrate ways of organizing the resources needed to complete a design project on time and within budget.

Webster's dictionary defines an *objective* as "something toward which effort is directed: an aim, goal, or end of action." Similarly, a *constraint* is a "condition, agency, or

force" that imposes a "stricture, restriction, or limitation." Thus, if we were designing a corn degrainer for Nicaraguan farmers to be cheaply built of indigenous materials, the objective might be that it should be as cheap as possible, and the constraint might be that it could not cost more than $20.00 (U.S.). Making the degrainer of indigenous materials could be an objective—is it a desired feature—or a constraint—a requirement.

In more colloquial terms our definition of engineering design would be:

Engineering design is the organized, thoughtful development and testing of characteristics of new objects that have a particular configuration or perform some desired function(s) that meets our aims without violating any specified limitations.

1.2.2   The assumptions behind our definition of engineering design

There are some implicit assumptions behind our two definitions of design and the terms in which they're written. It will be very helpful to make them explicit.

First, design is a *thoughtful* process that can be *understood*. Without meaning to spoil the magic or the importance of creativity in design, people do *think* while designing. We want to provide tools to support that thinking, to support design decision making and design project management. We know that people think while they design because we have written computer programs that simulate design processes. We couldn't do that if we couldn't articulate and describe what goes on in our heads when we design things.

The idea that there are *formal methods* to use when generating design alternatives is strongly related to our inclination to think about design. This might seem pretty obvious because there's not much point in looking for new ways of looking at design problems or talking about them unless we can exploit them to do design more effectively.

Form and *function* are two related yet independent entities. This is important. We often think of the design process as beginning when we sit down to draw or sketch something, which suggests that form is a typical starting point. However, we need to keep in mind that function is an altogether different aspect of a design that may not have an obvious relationship to its shape or form. In particular, while we can often infer the purpose of an object or artifact from its form or structure, we can't do the reverse, that is, we cannot automatically deduce from the function alone what form an artifact must have. For example, we can look at a pair of connected boards and deduce that the devices that connect them (e.g., nails, nuts and bolts, rivets, screws, etc.) are fastening devices whose function is to attach the individual members of each pair. However, if we were to start with a statement of purpose that we wish to connect two boards, there is no obvious link or inference that we can use to create a form or shape for a fastening device.

The relationship of form and function is important in understanding the creative aspects of design. If we can systematically articulate all of the functions that an airport is expected to perform, then we can be creative in developing forms within which these functions can be realized. In this sense, the use of organized, thoughtful processes *adds to* the creative side of design.

There are benchmarks used to specify in detail how we expect the design to perform, and they are also used to evaluate our progress toward a design. These benchmarks are

Engineering design is a *thoughtful* development of objects that perform desired functions within given limits.

Design is a thoughtful process that can be understood.

We cannot deduce form from function.

design specifications or *requirements* that result from a process (see Chapter 2) that begins with the designer

- *translating* the client's objectives into goals for the artifact being designed;
- *identifying* the means by which the design goals can be implemented;
- *delineating* the functions needed to achieve those means; and
- *specifying* the metrics that the design is *required* to achieve to demonstrate compliance, that is, the *design specifications*.

Design specifications are stated in a number of ways, depending on the nature of the requirements the designer wants to articulate. As we detail further in Chapter 5, design specifications may prescribe:

- *values* for particular design features;
- *procedures* with which attribute performance will be calculated; or
- *performance levels* that must be attained by the design.

Fabrication specifications enable fabrication independent of the designer's involvement.

The endpoint of a successful design is the production of a set of plans for making the designed artifact. This set of plans, called *fabrication specifications*, must be clear, unambiguous, complete, and transparent. That is, the fabrication specifications must, by themselves, enable someone totally apart from the designer or the design process to make what the designer intended so that it performs as the designer intended. This is a facet of modern engineering practice that represents a departure from a time when designers were often craftsmen who made what they designed. These designer-fabricators likely allowed themselves latitude or shorthand in their design plans because they, as fabricators, knew exactly what they, as designers, intended. It has become rare in more recent times for engineers to make what they design. Rather, designs were typically "thrown over the wall" to a manufacturing department or to a contractor—and all the fabricator or contractor knew was "what's in the specs." However, current practice has evolved with regard to this issue, and we will discuss that in the next section.

It often happens that the manufacture or use of a device points up deficiencies that were not anticipated in the original design. Successful designs often produce unanticipated secondary or tertiary effects that can become *ex post facto* evaluation criteria. For example, because of its contribution to *unintended consequences* such as air pollution and traffic congestion, some regard the automobile as a failure. On the other hand, the automobile provides the personal transportation that it was intended to. Further, changing societal expectations have dictated serious redesign of many of its attributes.

Communication is a key issue in design.

Finally, our definition of engineering design and the related assumptions we have identified clearly rely heavily on the fact that communication is central to the design process. Some set of languages or representations is inherently and unavoidably involved in every part of the design process. From the original communication of a design problem, through the specification of requirements and its fabrication, the artifact being designed must be described and "talked about" in many, many ways. Thus, *communication is the key issue*. It is not that problem solving and evaluation are less important; they are extremely important. But they too must be done at levels and in styles—whether spoken or written languages, numbers, equations, rules, charts, or pictures—that are appropriate to the immediate task at hand. Successful work in design is inextricably bound up with the ability to communicate.

1.3   MORE ON DESIGN AND ENGINEERING DESIGN

People have been doing design from time immemorial. People have also been talking and writing about design for a long time, but for far less time than they've actually been designing things. There's much to be learned from that, including thoughts on why design is hard, as well as suggestions that learning to do design is like learning to dance or ride a bike. First, however, let's look at the evolution of design over time.

1.3.1   The evolution of design practice and thinking

Thinking back to early elementary artifacts, it is almost certainly true that the "designing" was inextricably linked with the "making" of these primitive implements. We have no record of a separate, discernible modeling process, but can we ever know for sure? Who can say that small flint knives were not consciously used as models for larger, more elaborate cutting instruments? The inadequacy of small knives for cutting into the hides and innards of larger animals could surely have been a logical driver for making a small flint knife larger. People must have *thought* about what they were making, recognizing shortcomings or failures of devices already in use and making more sophisticated versions.

But we really have no idea of *how* these early designers thought about their work, what kinds of languages or images they used to process their thoughts about design, or what mental models they may have used to assess function or judge form. If we can be sure of anything, it is that much of what they did was done by trial and error. (Nowadays, when trial solutions are generated by unspecified means and tested to eliminate error, we call it *generate and test*.)

We find examples of works that must have been designed, such as the Great Pyramids of Egypt, the cities and temples of Mayan civilization, and the Great Wall of China far back in recorded history. Unfortunately, the designers of these wonderfully complex structures did not leave a "paper trail" that recorded their thoughts on their designs. However, there are some discussions of design that go pretty far back, one of the most famous being the collection of works by the Venetian architect Andrea Palladio (1508–1580). His works were apparently first translated into English in the eighteenth century. Since then, discussions of design have been developed in fields as diverse as architecture, organizational decision making, and various styles of professional consultation, including the practice of engineering. That's one reason there are a lot of definitions of engineering design.

Even in ancient times, designers evolved from pragmatists who likely designed artifacts as they made them, to more sophisticated practitioners who created sometime immense artifacts that others constructed. It has been said that the former approach to design, wherein the designer actually produces the artifact directly, is a distinguishing feature of a *craft*, and is found in such modern and sophisticated endeavors as graphics and type design.

1.3.2   The evolution of engineering design

We might similarly distinguish engineering design from other design domains or crafts by noting that engineering designers do not, typically, produce artifacts; rather, they produce the fabrication specifications for making artifacts. The designer in an engineering context produces a detailed description of the designed device so that it can be assembled or manufactured, thus separating the "designing" from the "making." This specification must be both complete and quite specific; there should be no ambiguity and nothing can be left out.

Traditionally, fabrication specifications were presented in a combination of drawings (e.g., blueprints, circuit diagrams, flow charts, etc.) and texts (e.g., parts lists, materials specifications, assembly instructions, etc.). Completeness and specificity can be achieved with such traditional specifications, but may not capture the designer's intent—and this can lead to catastrophe. In 1981, a suspended walkway in the Hyatt Regency Hotel in Kansas City collapsed because a contractor fabricated the connections for the walkways in a manner different than intended by the original designer.

In that design, walkways at the second and fourth floors were hung from the same set of threaded rods that would carry their weights and loads to a roof truss (see Figure 1.4). The fabricator was unable to procure threaded rods sufficiently long (i.e., 24 ft.) to suspend the second-floor walkway from the roof truss, so instead, he hung it from the fourth-floor walkway with shorter rods. (It also would have been hard to screw on the bolts over such lengths and attach walkway support beams.) The fabricator's redesign was akin to requiring that the lower of two people hanging independently from the same rope change his position so that he was grasping the feet of the person above, and that person was then carrying both their weights with respect to the rope. In the hotel, the supports of the fourth-floor walkway were not designed to carry the second-floor walkway in addition to its own dead and live loads, so a collapse occurred, 114 people died, and millions of dollars of damage was sustained. If the fabricator had understood the designer's intention to hang the second-floor walkway directly from the roof truss, this accident might never have happened. Had the designer explicitly communicated his intentions to the fabricator, a great tragedy might have been avoided.

There's another lesson to be learned from the separation of the "making" from the "designing." If the designer had worked with a fabricator or a supplier of threaded rods while he was designing, he would have learned that no one made threaded rod in the lengths

> The design is the only connection between the designer and the fabricator.

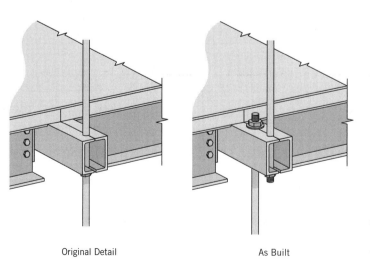

Original Detail   As Built

FIGURE 1.4 The walkway suspension connection as originally designed and as built in the Regency Hyatt House in Kansas City. We see that the change made during construction resulted in the second floor walkway being hung from the fourth floor walkway, rather than being connected directly to the roof truss.

needed to hang the second-floor walkway directly from the roof truss. Then the designer could have sought another solution in an early design stage. In much of the manufacturing and construction business, it was typically the case that there was a "brick wall" between the design engineers and the manufacturing engineers and fabricators. Only recently has this wall been penetrated. Manufacturing and assembly considerations are increasingly addressed during the design process, rather than afterward. One element in this new practice is *design for manufacturing*, in which the ability to make or fabricate an artifact is specifically incorporated into the design specifications, perhaps as a set of manufacturing constraints. The idea here is that the designer must be aware of parts that are difficult to make or of limitations on manufacturing processes as a design unfolds.

A second new idea is *concurrent engineering*, wherein designers, manufacturing specialists, and designers concerned with the product's life cycle (e.g., purchase, support, use, and maintenance) work together, along with others who hold a stake in the design, so that they are collectively and concurrently designing an artifact together. Concurrent engineering demands teamwork of a high order. Research in this area focuses on ways to enable teams to work together on complex design tasks when team members are dispersed not only by engineering discipline, but also geographically and by time zones.

The Hyatt Regency tale and the lessons drawn suggest that the statement or representation of fabrication specifications is very important. We learned that it is essential that fabrication specifications be complete and unambiguous. The designer's intent must be clearly communicated. Further, these specifications provide a basis for evaluating how well a design meets its original design goals. Thus, a designer wants to be able to translate the original objectives (and constraints) of the client into specifications, and recognize that these same requirements provide the starting point for manufacturing and design evaluation.

But while worrying about specifications, as well as all the other issues raised, keep in mind that design is a human activity, a social process. This means that communication among and between stakeholders remains a preeminent concern.

> The designer's intent must be clearly communicated to the fabricator.

1.3.3 A systems-oriented definition of design

There are a lot of words that appear often in discussions of design, for example, form, function, specification, and optimum. One author observed, however, that in one list of eleven definitions of design, only about one-tenth of the "really important words" appeared more than once in that entire list! Thus, the many, many definitions of design sound like they're trying to say similar things but in different words. This is because it is as hard to define design generally as it is to define the particular endeavor of engineering design. For example, we might define design as a goal-directed activity, performed by humans, and subject to constraints. The product of this design activity is a *plan* to realize those goals.

Herbert A. Simon, late Nobel laureate in economics and founding father to several fields, including design theory, offered a definition that is closely related to our engineering concerns and at the same time offers a broad perspective:

> As an activity, design is intended to produce a "description of an artifice in terms of its organization and functioning—its interface between inner and outer environments."

Designers are thus expected to describe the shape and configuration of a device (its "organization"), how that device does what it was intended to do (its "function"), and how the device (its "inner environment") works ("interfaces") within its operating ("outer") environment. Simon's definition is interesting for engineers because it places designed objects in a *systems* context, wherein we recognize that an artifact must operate as part of a system that includes the world around it.

1.3.4   Engineering design addresses hard problems

Engineering design problems are generally hard. They are hard because engineering design problems are usually *ill-structured* and *open-ended*:

Engineering design is an ill-structured, open-ended activity.

- Design problems are *ill-structured* because their solutions cannot normally be found by applying mathematical formulas or algorithms in a routine or structured way. Mathematics is both useful and essential in engineering design, but much less so in the early stages when "formulas" are both unavailable and inapplicable.

- Design problems are *open-ended* because they usually have several acceptable solutions. Uniqueness, so important in many mathematics and analysis problems, simply does not apply to design solutions. In fact, more often than not designers work to reduce or bound the number of design options they consider lest they be overwhelmed by the possibilities.

Evidence for these two characterizations can be seen in the familiar ladder. Several ladders are shown in Figure 1.5, including a stepladder, an extension ladder, and a rope ladder. If we want to design a ladder, we can't identify a particular ladder type to target unless and until we determine a specific set of uses for that ladder. Even if we decide that a particular form is appropriate, say a stepladder for the household handyman, other questions arise: Should the ladder be made of wood, aluminum, plastic, or a composite material? How much should it cost? And, which ladder design would be the *best*? Can we identify the *best* ladder design, the *optimal design*? The answer is, "No," we can't stipulate a single or universally optimal ladder design.

How do we talk about some of the design issues, for example, purpose, intended use, materials, cost, and possibly other concerns? In other words, how do we articulate the choices and the constraints for the ladder's form and function? There are different ways of representing these differing characteristics by using various "languages" or representations. But even the simple ladder design problem becomes a complex study that shows how the two characteristics of poorly defined endpoints (e.g., what kind of ladder?) and ill-defined structure (e.g., is there a formula for ladders?) make design a tantalizing yet difficult subject. How much more complicated and interesting are projects to design a new automobile, a skyscraper, or a way to land a person on the moon.

1.3.5   Learning design by doing

For someone who wants to learn *how to do it*, design is not all that easy to grasp. Like riding a bicycle or throwing a ball, like drawing and painting and dancing, it often seems easier to say to a student, "Watch what I'm doing and then try to do the same thing yourself." There is a *studio* aspect to trying to teach any of these activities, an element of *learning by doing*.

One of the reasons that it isn't easy to teach someone how to do design—or to ride a bike or throw a ball or draw or dance—is that we are often better at *demonstrating* a skill

FIGURE 1.5 A collection of "devices that enable people to reach heights they would be otherwise unable to reach," i.e., ladders. Note the variety of ladders, from which we can infer that the design objectives involved a lot more than the simple idea of getting people up to some height. Why are these ladders so different?

than we are at *articulating* what we know about applying our various skills. Some of the skills just mentioned clearly involve some physical capabilities, but the difference of most interest to us is not simply that some people are more gifted physically than others. What is really interesting is that a softball pitcher cannot tell you just how much pressure she exerts when holding the ball, nor how fast her hand ought to be going, or in what direction, when she releases it. Yet, somehow, and almost by magic, the softball goes where it's supposed to go and winds up in the hands of a catcher. The real point is that the thrower's nervous system contains some knowledge that allows her to assess distances and choose muscle contractions to produce a desired trajectory. While we can model that trajectory, given initial position and velocity, we do not yet have the ability to model the knowledge in the nervous system that generates that data.

Design is best learned by both doing and studying.

Note also that designers, like dancers and athletes, *use drills and exercises* to perfect their skills, *rely on coaches* to help them improve both the mechanical and interpretive aspects of their work, and *pay close attention* to other skilled practitioners of their art. Indeed, one of the highest compliments paid to an athlete is to say that he or she is "a student of the game."

1.4 MANAGING ENGINEERING DESIGN

Good design doesn't just happen.

Good design isn't something that just happens. Rather, it results from careful thought about what clients and users want and demand, and the specification of ways to realize those requirements. That is why the coming chapters will focus on the various tools and techniques that assist the designer in this process. One particularly important element of doing good design is *managing* the design project. Just as thinking about design in a rigorous way doesn't imply any loss in creativity, using tools to manage the design process doesn't mean sacrificing technical competency or inventiveness. On the contrary, there are many organizations that foster imaginative engineering design as an integral part of their management style. At 3M, for example, each of the more than 90 product divisions is expected to generate 25 percent of its annual revenues from products that didn't even exist five years earlier. Thus, a few tools and techniques of management that are applicable to design projects will also be introduced.

Just as we began by defining terms and developing a common vocabulary for design, we will do the same for management, project management, and the management of design projects. We will also go from the general to the specific in these definitions. Later we will look less at definition and more at "hands-on" matters. For now, we define management as follows:

Management achieves organizational goals by planning, organizing, leading, and controlling.

Management *is the process of achieving organizational goals by engaging in the four major functions of planning, organizing, leading, and controlling.*

This definition emphasizes that organizational goals are not achieved without certain processes. In this sense, management has something in common with design insofar as both are goal-directed and can be considered in terms of steps or processes. We will carry this analogy a little further in Chapter 2 when we consider the phases or stages of design.

The four functions of management can also be defined and discussed in ways that help see how management might relate to design.

- *Planning* is "the process of setting goals and deciding how best to achieve them." Planning involves considering the mission of the organization and translating that aim into appropriate strategic and tactical goals and objectives for the organization.

- *Organizing* is "the process of allocating and arranging human and nonhuman resources so that plans can be carried out successfully." Put another way, the organizing function of management is concerned with "creating a framework for developing and assigning tasks, obtaining and allocating resources, and coordinating work activities to achieve goals."

- *Leading* is the ongoing activity of exerting influence and using power to motivate others to work toward reaching organizational goals. That leadership results from influence is very important in design settings, where a number of different types of influence can come into play. For example, one member of the design team may have influence because of position (e.g., the team leader), while another may have influence based on the team's recognition of her expertise in a particular domain.

- *Controlling* is the process of monitoring and regulating the organization's progress toward achieving goals. Many people confuse leading and controlling, perhaps because we use the term "controlling" as a less-than-flattering synonym for using power in some settings. For engineers it is more complex because we use the term "control" to refer to directing system performance by monitoring and regulating. In this book, control refers to ensuring that the actual performance conforms to expected standards and goals.

Project management is the application of these four functions to accomplish the goals and objectives of a project. A project is "a one-time activity with a well-defined set of desired end results." Examples of engineering projects abound, ranging from constructing new highways (civil engineering) to developing new computer memories (electrical engineering), to establishing processes to be followed on a factory floor (industrial engineering). The common thread in these three projects is that each can be well defined in terms of its goals, has finite resources, and is to be accomplished in a fixed time frame (sometimes simply as soon as possible). To help project managers in carrying out the four functions (i.e., planning, organizing, leading, and controlling), a number of tools and techniques have been developed. These include tools for understanding and listing the work to be done, scheduling the tasks to be done logically and efficiently, assigning tasks to individuals, and monitoring progress. We will explore some of these tools and techniques as they are most applicable to design projects later in the book.

Note that the precision in goals and objectives that is spoken of regarding projects is somewhat at odds with some of the previous discussion of the open-ended nature of design activities. This is certainly the case when we try to predict the final form or outcome of a design project. Unlike a construction project, where the desired and expected results are clear and generally well articulated, a design project, and especially a conceptual design project, may have a number of possible successful outcomes, or none! This makes the task and tools of project management only partially useful in design settings. As a result, we will only present project management tools that we have found to be useful in managing design projects conducted by small teams.

In addition to a more restricted set of tools for managing the design project, we will introduce formal tools for guiding the design process itself. These tools are also a form of project management as they help the team to understand and agree upon goals, organize their activities, organize resources to realize the goals, and monitor whether the alternatives they generate and ultimately select are consistent with their objectives.

1.5 NOTES

Section 1.2: Our preferred definition of engineering design appears in (Dym and Levitt, 1991) and (Dym, 1994a).

Section 1.3: Simon's definition of design is based on a set of lectures that were published as *The Sciences of the Artificial* (1996).

Section 1.4: Definitions of management and the four major functions of planning, organizing, leading, and controlling are found in (Bartol and Martin, 1994) and (Bovee et al., 1993). The project is defined in (Meredith and Mantel, 1995).

1.6 EXERCISES

1.1 Two definitions of engineering design have been offered in this chapter. Identify the key concept in each definition and explain how these two concepts differ.

1.2 List at least three questions you would ask if you were, respectively, a user (purchaser), a client (manufacturer), or a designer who was about to undertake the design of a portable electric guitar.

1.3 List at least three questions you would ask if you were, respectively, a user (purchaser), a client (manufacturer), or a designer who was about to undertake the design of a greenhouse for a tropical climate.

1.4 All aspects of management may be said to be goal directed. Explain how this description is exemplified for each of the four functions of management identified in Section 1.4.

CHAPTER 2

THE DESIGN PROCESS

Can you give me a roadmap of where this book is going?

HAVING DEFINED engineering design and its attendant vocabulary, we go on to explore design as an activity, that is, the *process* of design. Some of this may seem abstract because we are trying to describe a very complex process by breaking it down into smaller, more detailed pieces. To anchor the tools we will describe in Chapters 3–5, we will identify those places in the design process where particular tools or methods can be useful. Please keep in mind, however, that *we are not presenting a recipe* for completing a design. Rather, we are trying to describe *what is going on in our heads* when we are doing design.

2.1 HOW A DESIGN PROCESS UNFOLDS

Suppose we are asked to design a safe ladder. Many safe ladders have already been designed, produced, and sold—a few of which we see in this chapter. So, what does it mean to start another "safe ladder" project?

2.1.1 Questions about the "safe ladder" project

There are many questions that arise as we think about the design of a safe ladder. A short, partial list could include:

- How is the ladder to be used?
- How much should it cost?
- What does "safe" mean?
- What is the market for this ladder?
- How many steps are there on the ladder?
- Is this design economically feasible?
- Is the ladder safe as it is actually designed?

Other questions in this simple design problem will soon appear below. Asking questions to elaborate the meaning of a client's statement is a part of the design process that is often done by design teams. It is important to think about the questions we need to ask, while at the same time asking who else in the designer-client-user triangle may have useful questions (and answers).

Some of the questions about the design above can be answered by applying the usual mathematical models of physics. Newton's equilibrium law and elementary statics could be used to analyze the stability of the ladder under given loads on a specified surface. We could write beam equations to calculate deflections and stresses in the steps as they bend under the given foot loads. But there are no equations that define the meaning of "safe," or of marketability, or that help us choose a color for the ladder. Since there are no equations for safety, color, marketability, or for most of the other issues in the ladder questions, we must find other ways to think about this design problem.

There are no equations for safety, color, marketability . . .

We can use such lists of questions (and their answers) to get a handle on the design process and to think about the tasks we are doing when we ask those questions. We find that we can *decompose* or break down the process into a sequence of steps (or design tasks) by extracting and naming those steps.

Questions such as:

- How is the ladder to be used?
- How much should it cost?

help us to *clarify the objectives* set for the design by the client.

Questions such as:

- Should the ladder be portable?
- How much can it cost?

help us to *establish user requirements* for the design.

Questions such as:

- How is safety defined?
- What is the *most* that the client is willing to spend?

help us to *identify constraints* that govern the design.

Questions such as:

- Can the ladder lean against a supporting surface?
- Must the ladder support someone carrying something?

help us to *establish functions* for the design.

Questions such as:

- How much weight should a safe ladder support?
- What is the "allowable load" on a step?
- How high should someone on the ladder be able to reach?

help us to *establish design specifications* for the design.

Questions such as:

- Could the ladder be a stepladder or an extension ladder?
- Could the ladder be made of wood, aluminum, or fiberglass?

help us to *generate design alternatives*.

Questions such as:

- What is the maximum stress in a step supporting the "design load"?
- How does the bending deflection of a loaded step vary with the material of which the step is made?

help us to *model* and *analyze* the design.

Questions such as:

- Can someone on the ladder reach the specified height?
- Does the ladder meet OSHA's safety specification?

help us to *test* and *evaluate* the design.

Questions such as:

- Is there a more economic design?
- Is there a more efficient design (e.g., less material)?

help us to *refine* and *optimize* the design.

Finally, questions such as:

- What information does the client need to fabricate the design?
- What is the justification for the design decisions that were made?

help us to *document* the completed design and the design process.

We see that the questions we are asking about the ladder design establish steps in a process that move us from an abstract statement of a design objective through increasing levels of detail as we: *build a model* of the ladder; *analyze and test* that model; *optimize and refine* some of its features; and we *document* both the fabrication specifications and the justification for this particular design. Thus, we identified some of the tasks that need to be done in order to complete a design. We will identify additional tasks, as well as describe in greater detail *all* of the tasks of engineering design.

2.1.2 Commentary on the questions about the "safe ladder" project

A design project typically begins with a verbal statement that states desired features of

- *function*;
- *form*;
- *intent*; or of
- *legal requirement*.

It is the designer's job to clarify what the client wants and then to translate the client's wishes into more concrete objectives. In the clarification step the client is asked to be more precise about what he really wants by asking questions such as: For what are you going to use the ladder? Where? How much can the ladder itself weigh? What level of quality do you want in this ladder? How do you define quality? How much are you willing to spend?

Some of the questions asked to clarify the client's wishes obviously connect with a later part of the process in which we make choices, analyze how possibly competing choices interact, assess any trade-offs in these choices, and evaluate the effect of these choices on our overall goal of designing a safe ladder. For example, the ladder's form or configuration is strongly related to its function: We are more likely to use an extension ladder to rescue a cat from a tree and a stepladder to paint the walls of a room. Similarly, the weight of the ladder has an impact on the efficiency with which it can be used: Aluminum extension ladders have replaced wooden ones largely because they weigh less. The material of which a ladder is made not only affects its weight, but also its cost and its feel: Wooden extension ladders are much stiffer than their aluminum counterparts, so users of the aluminum versions get used to feeling a certain amount of "give" or flex in the ladder, especially when it is significantly extended.

The next part of the design process is where the client's wishes are translated into a set of *user requirements* that state in great detail what is wanted from the design by the client and by potential users. These user requirements are detailed expressions of the designed object's desired functionality and of its attributes. These requirements are the basis for elaborating a set of *design specifications* that serve as benchmarks against which the performance of a designed artifact is measured. Design specifications are typically stated in one of three ways, depending on the nature of the requirements we want to articulate (see Chapter 5 for more on this point):

- *prescriptive specifications* specify values for attributes of the designed object;
- *procedural specifications* identify specific procedures for calculating attributes or behavior; and
- *performance specifications* characterize the desired behavior.

A successful design meets or exceeds the given design specifications.

We will come across a vast array of choices to make as a design evolves. At some point in our ladder design, for example, we will have to choose a *type* of ladder, say a stepladder or an extension ladder. We will have to decide how to fasten the steps to the ladder frame. Our choices will be influenced by the desired behavior (e.g., although the ladder itself may flex, we don't want individual steps to have much give with respect to the ladder frame) and by manufacturing or assembly considerations (e.g., would it be better to nail in the steps of a wooden ladder, use dowels and glue, or nuts and bolts?). Here we are decomposing the ladder into its components or pieces, and selecting particular types of components.

We should also note that as we work through these steps, we are constantly communicating with others about the ladder and its various features. When we question the client about its desired properties, for example, or the laboratory director about the evaluation tests, or the manufacturing engineer about the feasibility of making certain parts, we are interpreting aspects of the ladder design in terms of "languages" and parameters that these experts use in their own work. Thus, the design process can't proceed without such interpretations.

This simple design problem shows how we can *formalize* the design process to make explicit the design tasks that we are doing. We are also *externalizing* aspects of the process, moving them from our heads into some recognizable language(s) for communication to others, as well as for further use. Thus, we learn two important lessons from our ladder design project:

Clarifying objectives and translating them into the right "language" are essential elements of design.

- An essential part of an engineering design project is *clarifying* the client's objectives. The designer must fully understand what the client wants and the users need from the resulting design. Good communication among all three parties in the designer-client-user triangle is essential, and we must be very careful in eliciting the details of what the client really wants.

- The remaining tasks in the design process involve *translating* the client's objectives into the kinds of words, pictures, numbers, rules, properties, etc., that we need to characterize and describe both the object being designed and its behavior. The tasks of analyzing and modeling, testing and evaluating, and refining and optimizing, cannot be done verbally. Documenting the final design also cannot be done with words alone. We need pictures and numbers, and perhaps other ways of representing the desired design. Thus, the designer translates the client's verbal statement into whatever forms are appropriate to doing the various design tasks at hand.

2.2 DESCRIBING AND PRESCRIBING THE DESIGN PROCESS

There are many models of the design process, just as there are many definitions of design. Some design process models are *descriptive*, that is, they attempt only to *describe* the elements of the design process. Other models are *prescriptive* in that they *prescribe* what should be done during the design process.

2.2.1 Describing the design process

One of the simplest descriptive models of the design process has three stages:

- *generation,* in which the designer proposes various concepts that are generated or created;
- *evaluation,* during which the design is tested against criteria that have been set forth by the client, the user(s), and the designer; and
- *communication*, when the design is communicated to the manufacturers or fabricators.

A similar three-stage model calls for *doing research, creating,* and then *implementing* a final design, with the contexts providing meanings for these three steps. While these two models have the virtue of simplicity, they are so abstract that they provide little useful advice on how to do a design. To cite one very obvious question, *how* do we generate or produce designs?

Another, widely accepted model of the design process is portrayed in Figure 2.1, with its three "active" stages shown in boxes with rounded corners. The starting point for this descriptive model is the client's statement, which is often identified as the *need* for a design. The endpoint is the final design or the set of fabrication specifications. The first stage of the process depicted in Figure 2.1 is *conceptual design,* in which we look for different *concepts* (also called *schemes*) that can be used to achieve the client's objectives. A concept or scheme is an outline solution to a design problem. The means for achieving the major functions have been identified and fixed, as have the spatial and structural relationships of the

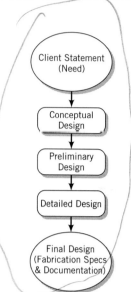

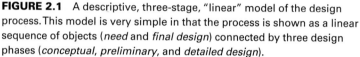

FIGURE 2.1   A descriptive, three-stage, "linear" model of the design process. This model is very simple in that the process is shown as a linear sequence of objects (*need* and *final design*) connected by three design phases (*conceptual*, *preliminary*, and *detailed design*).

principal components. Enough details have been worked out so that we can estimate costs, weights, and overall dimensions.

For the ladder project, we might have chosen an extension ladder, a stepladder, or even a rope as our concept. The evaluation of these schemes will depend on the attributes in the client's statement, for example, the intended use, specified cost, and even the client's aesthetic values.

With its focus on trade-offs between high-level objectives, conceptual design is clearly the most abstract and open-ended part of the design process. The output of the conceptual stage may be one, two, or several competing concepts. Some even argue that conceptual design *should* produce two or more schemes since early attachment to a single design choice may be a mistake. This tendency is so well known among designers that it has produced a saying: "Don't marry your first design idea."

The second phase in this model of the design process is *preliminary design* or, especially in Europe, *embodiment of schemes*. Here proposed schemes are "fleshed out," that is, we hang the meat of some preliminary choices upon the abstract bones of the conceptual design. We embody or endow design concepts with their most important attributes. We begin to select and size the major subsystems, based on lower-level concerns that take into account the performance specifications and the operating requirements. For the stepladder, for example, we size the side rails and the steps, and perhaps decide on how the steps are to be fastened to the side rails. Preliminary design is clearly more technical in nature, so we might use various back-of-the-envelope calculations. We make extensive use of rules of thumb about size, efficiency, and so on, that reflect the designer's experience. And, in this phase of the design process, we make our final choice from among the proposed concepts.

The final stage of this model is *detailed design*. We now refine the choices we made in the preliminary design, articulating our early choices in much greater detail, down to specific part types and dimensions. This phase typically follows design procedures that are

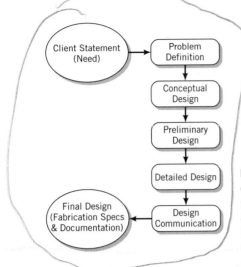

FIGURE 2.2 A five-stage, *descriptive* model of the design process. This model is very stylized in that the process is shown as a linear sequence of objects (*need* and *final design*) and five phases of the process design (*problem definition, conceptual design, preliminary design, detailed design,* and *design communication*).

quite well understood by experienced engineers. Relevant knowledge is found in design codes (e.g., the ASME Pressure Vessel and Piping Code, the Universal Building Code), handbooks, databases, and catalogs. Design knowledge is often expressed in specific rules, formulas, and algorithms. This stage of design is typically done by component specialists who use libraries of standard pieces.

The classic model just outlined can be extended to a five-stage model that delineates two additional sets of activities that precede and follow the three-stage model sequence:

- *problem definition* is a pre-processing stage that identifies the work done with the client's statement *before* conceptual design begins; and

- *design communication* is a post-processing phase that identifies the work done *after* detailed design to document and present the final design and fabrication specifications.

This five-stage model of the process, displayed in Figure 2.2, is more detailed than the three-stage models discussed above, but it does not bring us much closer to knowing *how* to do a design because it too is *descriptive*. In the next section we present a process model that prescribes what ought to done.

2.2.2 Prescribing the design process

In Figure 2.3 we show a *prescriptive* model of the design process that converts the five-stage descriptive model of Figure 2.2 into a prescriptive model that defines what is done in each stage by incorporating the ten design tasks identified in Section 2.1. Beginning with the client's statement and ending when the final design is documented for the client. each phase requires an *input*, has *design tasks* that must be performed, and produces an *output* or product. We now detail out the design tasks done in each stage, together with the sources of information, methods, and means. Note that the output of each stage serves as the input to the following stage.

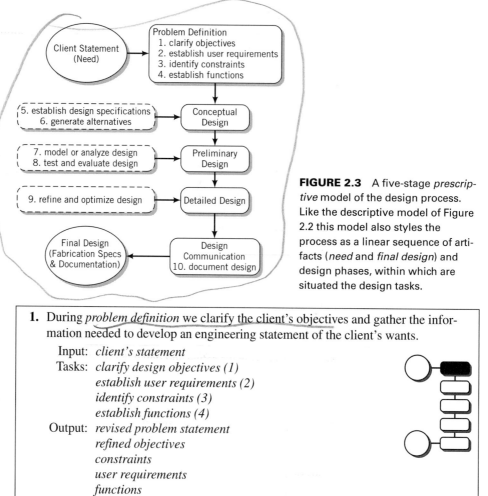

FIGURE 2.3 A five-stage *prescriptive* model of the design process. Like the descriptive model of Figure 2.2 this model also styles the process as a linear sequence of artifacts (*need* and *final design*) and design phases, within which are situated the design tasks.

1. During *problem definition* we clarify the client's objectives and gather the information needed to develop an engineering statement of the client's wants.

   Input: *client's statement*
   Tasks: *clarify design objectives (1)*
            establish user requirements (2)
            identify constraints (3)
            establish functions (4)
   Output: *revised problem statement*
            refined objectives
            constraints
            user requirements
            functions

The *sources of information* during problem definition include literature on state-of-the-art, experts, codes, and regulations. *Methods* include objectives trees, pairwise comparison charts, function-means trees, functional analysis, and requirements matrices. The *means* include literature reviews, brainstorming, user surveys and questionnaires, and structured interviews.

2. In the *conceptual design* stage of the design process we generate concepts or schemes of candidate designs.

   Input: *revised problem statement*
            refined objectives
            constraints
            user requirements
            functions
   Tasks: *establish design specifications (5)*
            generate design alternatives (6)
   Output: *conceptual design(s) or scheme(s)*
            design specifications

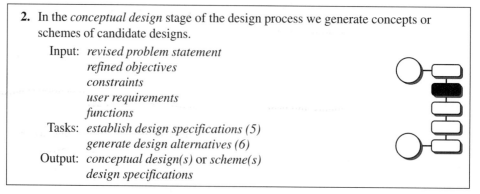

Competitive products are the additional principal *sources of information* for conceptual design. *Methods* include the performance specification method, quality function deployment (QFD), and morphological charts. *Means* include brainstorming, synectics and analogies, benchmarking, and reverse engineering (dissection).

---

3. In the *preliminary design* phase we identify the principal attributes of the design concepts or schemes.

    Input: *conceptual design(s) or scheme(s)*
              design specifications
    Tasks: *model, analyze conceptual designs (7)*
            test, evaluate conceptual designs (8)
    Output: *a selected design*
            test-and-evaluation results

The *sources of information* include heuristics (rules of thumb), simple models and known physical relationships. *Methods* include refined objectives trees and pairwise comparison charts. *Means* include metrics definition, laboratory experiments, prototype development, simulation and computer analysis, and proof-of-concept testing.

---

4. During *detailed design* we refine and detail the final design.

    Input: *selected design*
            test-and-evaluation results
    Task: *refine and optimize the chosen design (9)*
    Output: *proposed fabrication specifications*
            final design review for client

The *sources of information* for detailed design include design codes, handbooks, local laws and regulations, and suppliers' component specifications. *Methods* include discipline-specific CADD. *Means* include formal design reviews, public hearings (if applicable), and beta testing.

---

5. Finally, during the *design communication* phase we document the fabrication specifications and their justification.

    Input: *fabrication specifications*
    Task: *document the completed design (10)*
    Output: *final report to client containing*
            (1) fabrication specifications
            (2) justification for fabrication specs

The *sources of information* during the design communication stage are feedback from clients and users, and itemized lists of required deliverables.

We now have a "checklist" we can use to ensure that we have done all of the "required" steps. Lists like this are often used by design organizations to specify and propagate approaches to design within their firms. However, we should keep in mind that this and other detailed elaborations add to our understanding of the design process only in a limited way. At the heart of the matter is our ability to model the tasks done within each phase of the design process. With this in mind, we will present some means and formal methods for doing these ten design tasks in Section 2.3.

2.2.3   Feedback and iteration in the design process

All of the models we have presented so far have been "linear" or sequential. But the design process is not linear or sequential. We have intentionally left out two very important elements in design thinking. The first of these is *feedback*, that is, the process of feeding information about the output of a process back into the process so it can be used to obtain better results. Feedback occurs in two notable ways in the design process, as illustrated in Figure 2.4.

The first feedback mechanism is an internal *feedback loop* wherein the results of performing the test and evaluation task are fed back into the preliminary design stage to *verify* that the design performs as intended. And, as detailed in Section 1.3 in our discussion of concurrent engineering, there will be feedback from internal customers, such as manufacturing (e.g., can it be made?) and maintenance (e.g., can it be fixed?).

The second feedback mechanism is an external loop. It occurs after the final product resulting from a design has been used in the market for which it was intended. User feedback then provides *validation* for the design, assuming it is a successful design.

The second element that we have thus far left out of our process models is *iteration*. We iterate when we repeatedly apply a common method or technique at different points in a design process (or in analysis), although the repeated applications occur at a different level of abstraction or on a different scale. These repetitions typically occur at more refined, less abstract points in the design process (and on a finer, more detailed scale in analysis). In terms of the linear, five-stage model depicted in Figure 2.3, we should anticipate repeating the first four tasks in some form in conceptual, preliminary, and detailed design. That is, we

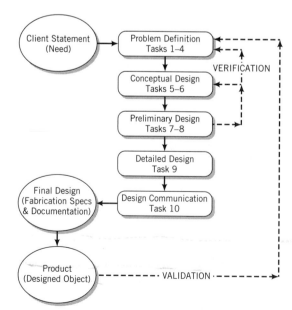

FIGURE 2.4   A five-stage prescriptive model of the design process that shows (in dashed lines) feedback loops for: (a) *verification*, internal feedback during the design process; and (b) *validation*, external feedback obtained from users when they actually use the designed object.

Iteration and feedback are integral parts of the design process.

always want to keep the original objectives in mind to ensure that we have not strayed from them as we get deeper and deeper into the details of our final design. Of course, this may also mean that we may have reason to do some redesign, in which case we will certainly reiterate tasks 7 (analyzing the design) and 8 (testing and evaluating the design) as well.

Given that there are feedback loops and that we will repeat or reiterate some tasks, why did we present our process models as linear sequences? The answer is simple. We noted in Chapter 1 that "design is a goal-directed activity, performed by humans." As important as the feedback and iterative elements of design are, it is equally important not to be overly distracted by these adaptive characteristics when learning about—and trying to do—design for the first time. It is also true that, in some sense, feedback loops and the need to repeat some design tasks will occur naturally as a design project unfolds. When we are doing a design for a client, it is only natural to go back and ask the client if the original project statement was properly revised. It is just as natural to show off emerging design concepts and to refine these schemes by responding to feedback, and by reiterating objectives, requirements, and specifications.

2.2.4 On opportunities and limits

The primary focus of this book is conceptual design, the first phase of the design process. As a result, we will often be dealing with some very broad themes and approaches in ways that are logical, but not as neat and tidy as a set of formulas or algorithms. In fact, conceptual design tools lend themselves to answering questions that are not easily posed in formal mathematical terms. It is ironic that this seeming lack of rigor of the tools we will present for use in conceptual design also makes them very useful more generally for *problem solving*.

At the same time, we should remember that we are limiting our detailed discussion of design methods to the earliest phases of design. We are not delving into the many processes and activities that must be done before a design can be turned into a product that a customer can actually buy.

2.3 STRATEGIES, METHODS, AND MEANS IN THE DESIGN PROCESS

Even prescriptive descriptions of design processes can fail us because they don't tell us *how* to generate or create designs. Here we will briefly introduce some of the formal design methods and some of the means of acquiring design-related information methods, as a prelude to the more detailed descriptions given in Chapters 3–5. Remember that we are introducing these decision-support techniques and tools to explain *how* we go about designing artifacts, that is, we are describing *thought processes* or *cognitive tasks* that will be done during the design process. We begin with ideas for strategic approaches to design thinking.

2.3.1 Strategic thinking in the design process

It is generally unwise to commit to a particular concept or configuration until forced to by the exhaustion of additional information or of alternate choices. This general strategy for thinking about design is called *least commitment*. (Remember that saying, "Never marry your first design.") Least commitment is less a method than it is a strategy or a good habit of thought. It militates against making decisions before there is a reason to make them. Premature commitments can be dangerous because we might become attached to a bad concept or we might limit ourselves to a suboptimal range of design choices. Least commitment is of particular importance in conceptual design because the consequences of any early design decision are likely to be propagated far down the line.

Another important strategy of design thinking is to apply the power of *decomposition*, that is, of breaking down, subdividing, or decomposing larger problems (or entities or ideas) into smaller subproblems (or subentities or subideas). These smaller subproblems are usually easier to solve or otherwise handle. That is why decomposition is often identified as *divide and conquer*. We have to keep in mind that subproblems can interact, so we must ensure that the solutions to particular subproblems do not violate the assumptions or constraints of complementary subproblems.

2.3.2 Some formal methods for the design process

We now present brief introductions to the formal design methods that are listed in Figures 2.1–5 for the five stages of the design process.

We build *objectives trees* in order to clarify and better understand a client's project statement. Objectives trees are hierarchical lists that branch out into tree-like structures and in which the objectives that designs must serve are clustered by sub-objectives and then ordered by degrees of further detail. The highest level of abstraction of an objectives tree is the top-level design goal, derived from the client's project statement. In Section 3.1 we detail how we construct objectives trees and what kind of information we learn from them.

We rank design objectives using *pairwise comparison charts*, a relatively simple device in which we list the objectives as both rows and columns in a matrix or chart and then compare them on a pair-by-pair basis, proceeding in a row-by-row fashion. The pairwise comparison chart is useful for rank ordering objectives early in the design process. It is also helpful for choosing among competing attributes or requirements. We describe pairwise comparison charts in Section 3.3.

Functional analysis is used to identify what a design must do. The starting point for analyzing the functionality of a proposed device is usually a "black box" with a clearly delineated boundary between the device and its surroundings. In functional analysis we often decompose overall function(s) into subfunction(s). This is achieved by tracking the flow of materials, signals, etc., through the device and detailing the material- or signal-processing needed to produce the desired functions. We present functional analysis and the related *function-means tree* in Section 4.1.

The *performance specification method* provides support for the elaboration of the design specifications that are the designer's target for a design project. The aim is to list solution-independent attributes and performance specifications (i.e., "hard numbers") for both required and desired features of a design concept. We describe performance specifications and their use in Section 5.2.

The *morphological chart* is used to identify the ways or means that can be used to make the required function(s) happen. A "morph" chart provides a framework for visualizing a *design space*, that is, an imaginary "plane," "room," or "space" that we can use to generate, collect, identify, store, and explore all of the potential design alternatives that might solve our design problem. We describe design spaces more formally in Section 5.1 and morph charts in Section 5.3.

A more advanced tool, *quality function deployment* (QFD), builds on the performance specification method with the goal of achieving higher-quality products. Used widely in product manufacturing, QFD calls for charting client and user *requirements* and engineering attributes in a *matrix* format that makes it possible to relate and weigh them, one against another. The intent is to erect a *house of quality* that exposes both positive and negative interactions of the engineering specifications, thus enabling a designer to anticipate and weed out performance conflicts. QFD is briefly described in Section 8.5.

2.3.3 Some means for acquiring and processing design knowledge

Here we describe the means by which information is gathered and analyzed for ultimate use in the formal design methods. These means are tools that have been developed in a number of disciplines. We organize them into three categories: means for acquiring information; means for analyzing the information obtained and testing outcomes against desired results; and means for obtaining feedback from clients, users, and stakeholders (i.e., other interested parties). In many cases the means are so widely applied that we view their detailed description as beyond our scope. In other cases, however, the means are so important that we will provide expanded discussions later in the book.

2.3.3.1 Means for acquiring information The classic method of determining the state of the art and prior work in the field, the *literature review*, is so familiar that it might seem superfluous for us to comment on it. In the preprocessing and conceptual stages we need literature reviews to enhance our understanding of the nature of potential users, the client, and the design problem itself. We are then able to consider prior or existing solutions, including product advertising and vendor literature. In preliminary design we are likely to look more toward the technical literature regarding the physical properties of possible solutions. In the detailed design stage we will want to look at handbooks, compendia of material properties, design and legal codes, etc.

User surveys and questionnaires are used in market research. They naturally focus on identifying user understanding of the problem space and user response to possible solutions. Market research can help the designer to clarify and better understand the problem itself in its early stages, so that questions are necessarily open-ended. Later surveys may be used together with pairwise comparison and morphological charts to assist in selection and choice.

Focus groups are an expensive way of allowing a design team to observe the response of appropriately selected users and others to potential designs. Since the intelligent use of focus groups demands considerable sophistication in matters psychological, they are not generally used by student design teams.

On the other hand, *informal interviews* are often undertaken very early in a design project, when the team is still trying to define the problem sufficiently to plan an approach. While informal interviews are relatively simple to conduct, it is important to be sensitive to the time and other constraints of the interviewee. All too often a design team will simply show up and ask seemingly random questions, with the dual effect of highlighting their ignorance while doing little or nothing to dispel it. There are ways to reduce this problem, including sending the interviewee copies of the topics and questions in advance and undertaking extensive literature research prior to conducting interviews.

The *structured interview* is another means of eliciting information that combines elements of a survey form with the flexibility of informal interviews. Here the interviewer uses a previously defined set of questions that may or may not be made available to the interviewees. In addition to using the question set to get direct answers, the interviewer can follow up a particular response and open up new areas. The structured set of questions also assures the person being interviewed that the interview has both purpose and focus, and it ensures that interesting side issues do not prevent the key matters from being covered.

Another way that design teams can acquire further insight is to do some *brainstorming*, an activity that allows the participants to generate related or even unrelated ideas that are listed but not evaluated until a later time. The free-wheeling nature of brainstorming can be very helpful in opening up new avenues for research and analysis. However, as we will note in a more detailed discussion of brainstorming in Section 6.2, it is very important that team members maintain a high level of respect for the ideas of others and that *all* ideas are captured and listed as they are offered.

Design teams can draw upon their own abilities to discover and explore relationships and similarities between ideas and solutions that initially seem unrelated. In an environment free of criticism and evaluation, similar to that in brainstorming, the team conducts a *synectic* activity in which it tries to uncover or develop *analogies* between one type of problem and other types of problems or phenomena. For example, a team might seek to redefine a problem into terms of some other matter solved by an earlier team. Or, the team might initially seek to find the most outrageous solutions to a problem and then look for ways that such solutions could be made useful. We will say more about synectics in Section 5.2, but we note that it is time consuming and demands serious commitment on the part of the design team. Consequently, synectics is generally more widely used in industrial design settings than in academia.

Finally, design teams often use two (unrelated) activities to look at "what's out there" when developing new products. In one case, *competitive products* are *benchmarked*, that is, designers literally look at similar products that are already available and try to evaluate how well those products perform certain functions or exhibit certain features. These competitive products serve as standards for "building a better mousetrap" since proposed new designs can be compared to existing mousetraps. The second activity, *reverse engineering* or *dissection*, consists of dissecting or (literally) taking competitive or similar or prior products apart. The idea is to determine why a given product or device was designed the way it was, with the intention of finding better ways of performing the same or similar subfunctions.

2.3.3.2 Means for analyzing information and testing outcomes   An important first step in determining whether or not design concepts might work is to *define the metrics* against which outcomes can be measured. Of particular interest is the relationship between the expression of a design's stated objectives and constraints in the design specifications on the one hand, and the statement of acceptable and desirable design outcomes on the other. This is an extremely important topic that we will discuss in depth in Sections 4.4 and 5.3.

In some cases we find it possible to gather and assess data about a potential design solution by undertaking *laboratory experiments*. For example, if the solution involves a structure, it may be possible to measure the stress or strength relationships of critical parts of the design in a laboratory test.

One very important means of determining whether or not a design can perform its required functions is *prototype development*. Here a prototype or test unit may embody the principal functional characteristics of the final design, even though it may not look at all like the expected end product. In fact, early prototypes typically have only a subset of the required functionality to shorten development time and reduce costs. Prototypes can be instrumented to support laboratory or other tests.

A crucial step along the path from conceptual design to detailed design is *proof-of-concept testing*. Such testing involves establishing a formal means of determining whether or not the concept under consideration can reasonably be expected to meet the design requirements. In many cases a proof-of-concept test will not require that the tested unit survive or even perform the stated function(s). Rather, the idea is to show that a design can be made to fulfill the functions under certain prespecified conditions. As with any scientific test, we must define outcomes that are sufficient for accepting or rejecting a concept.

In many cases we can't develop or test a prototype, perhaps because of cost, size, hazard, or other reasons. In such cases we often resort to *simulation* wherein we exercise an analytical, computer, or physical model of a proposed design to simulate its performance under a stated set of conditions. This presumes that the device being modeled, the conditions under which it operates, and the effects of the operations are sufficiently well understood to make the model useful. One outstanding example of such simulation is the use of wind tunnels and related computer analyses to assess the effects of wind loading on tall buildings and long, slender suspension bridges.

Closely related to simulation, *computer analysis* involves the development of a computer-based model which may consist of the equations relevant to describing the design, and application of various analytic, discipline-based techniques. These include finite element analysis, integrated circuit modeling, failure mode analysis, criticality analysis, etc. Computer-based models are used widely in all engineering disciplines and they become even more important as a design project moves into detailed design. That's why we refer to such tools as *discipline-specific CADD* systems after Box 4 on page 25, or, discipline-specific computer-aided design and drafting systems.

2.3.3.3 Means for obtaining feedback

Among the most important means for obtaining feedback from both clients and users are *regularly scheduled meetings* at which the progress of the design project, including articulation of the various stages of the design process, is tracked and discussed. We assume throughout our discussions that the design team is always communicating with both clients and users. We will often suggest that various formal design results be reviewed with them.

At certain stages of the design process it is a standard practice to hold a *formal design review* wherein the current design is presented to the client(s), selected users, and/or other stakeholders. These presentations typically include sufficient technical detail that the implications of the design can be fairly explored and assessed. It is particularly important that young designers become comfortable with the "give and take" that often accompanies such reviews. While it may seem harsh to be asked to justify various technical details to clients and others, the design review process is usually beneficial because implicit unwarranted assumptions and errors or oversights are often uncovered.

In some design environments, relevant civil laws or public policies require that *public hearings* be held for the purpose of exposing the design to public review and comment. While it is beyond our scope to consider such hearings in detail, it is useful for designers to understand that just as teams are increasingly the internal organizational structure of choice for design, public hearings and meetings are increasingly the norm for major design projects, even when the client is a private corporation.

We have already noted that *focus groups* are important sources of user input on problem definition. Such groups are also used to assess user reaction to designs as they near adoption and marketing.

In some industries, most notably software design, an almost-but-not-quite-finished version of a product is released to a small number of users. This practice, called *beta testing*, allows the designers to uncover design or implementation errors and to get feedback about their product before it reaches a larger market. Beta testing will not be discussed further in this text.

2.4   GETTING STARTED: MANAGING THE DESIGN PROCESS

Just as there are many models describing the design process, there are also many depictions of project management. We show a brief road map of the project management path for design projects in Figure 2.5, in direct analog with Figure 2.3 for the design process. This figure highlights the fact that project management follows a path that encompasses:

- *project definition*—developing an initial understanding of the design problem and its associated project;

- *project framework*—developing and applying a plan to do the design project;

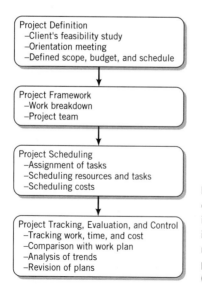

FIGURE 2.5   Managing a design project follows an orderly process, beginning with the client's understanding of the problem. At the early stages, the design team is concerned with understanding the problem and making plans to solve it. Later, the focus shifts toward project control and staying on plan. Adapted from (Orberlander, 1993).

- *project scheduling*—organizing that plan in light of time and other resource constraints; and as the project unfolds,
- *project tracking*—evaluation and control; keeping track of time, work, and cost.

In Chapter 7 we will detail a number of tools to help us proceed along this path. At this time, however, it is worthwhile to note how this model affects the team at the very beginning of a project. Many of the activities associated with project definition, such as the feasibility study, the orientation meeting, and setting the overall schedule and budget, may be beyond the control of the design team. Nevertheless, the team will usually devote its initial meetings and activities to trying to better understand these issues. One activity that cannot be deferred is the organization and development of the project team addressed in the next section.

2.4.1 Organizing design teams

Design is often done by teams rather than by individuals.

Design is an activity that is increasingly done by teams rather than individuals. For example, new products are often developed by teams that include designers, manufacturing engineers, and marketing experts. These teams are assembled to gather the diverse skills, experiences, and viewpoints needed to successfully design, manufacture, and sell new products. This dependence on teams is not surprising if we reflect on the stages, methods, and means for design that we have been discussing. Many of the activities and methods are devoted to applying different talents and skills to realize a common understanding of a problem. Consider, for example, the difference between laboratory testing and computer-based analysis of a structure. Both require common knowledge of structural mechanics, yet years of investment are required to master the specific testing and laboratory skills or the analysis and computer skills. Thus, there may be considerable value in constructing teams whose members have all the needed skills and can work together successfully. In this section we briefly introduce some aspects of team formation and performance, and we then relate them to one of the means for generating ideas discussed above, namely brainstorming.

2.4.1.1 Stages of group formation   Groups and teams are such an important element of human enterprise that we should not be surprised to learn that they have been extensively studied and modeled. One of the most useful models of group formation suggests that almost all groups typically undergo five stages of development that have been named as:

- *forming,*
- *storming,*
- *norming,*
- *performing,* and
- *adjourning.*

We will use this five-stage model to describe some of the elements of group dynamics that are often encountered in engineering design projects.

Forming: Most of us experience a number of feelings simultaneously when we are initially assigned to a team or group. These feelings range from excitement and anticipation to anxiety and concern. We may worry about our ability—or that of our teammates—to perform the tasks asked of us. We may be concerned about who will show the leadership needed to accomplish the job. We may be so eager to get started that we rush into assignments and activities before we are really ready to begin. Each of these feelings and concerns are elements of the *forming* stage of group development, characterized by a number of aspects and behaviors, including:

* becoming oriented to the (design) task at hand,
* becoming acquainted with the other members of the team,
* testing group behaviors in an attempt to determine if there are common viewpoints and values,
* being dependent upon whoever is believed to be "in charge" of the project or task, and
* attempting to define some initial ground rules, usually by reference to explicitly stated or externally imposed rules.

In this stage, the team members may often do or say things that reflect their uncertainties and anxieties. It is important to recognize this because judgments made in the forming stage may not prove to be valid over the life of a project.

Storming: After the initial or forming stage, most groups come to understand that they will have to take an active role in defining the project and the tasks needed to complete it. At this point the team may resist or even resent the assignment, and it may challenge established roles and norms. This period of group development is known as the *storming* phase and is often marked by intense conflict as team members decide for themselves where the leadership and power of the team will lie, and what roles they must individually play. At the same time, the team will usually be redefining the project and tasks, and discussing opinions about the directions the team should explore. Some characteristics of the storming phase are:

* resistance to task demands,
* interpersonal conflict,
* venting of disagreement, often without apparent resolution, and
* struggle for group leadership.

The storming phase is particularly important for the design team because there is often already a high level of uncertainty and ambiguity about client and user needs. Some team members may want to rush to solutions and will consider a more thoughtful exploration of the design space simply as stubbornness. At the same time, most design teams will not have as clear a leadership structure as, for example, a construction, manufacturing, or a research project. For these reasons, it is important for effective teams to recognize when the team is spending too long in the storming phase and to encourage all team members to move to the next phases, norming and performing.

Norming: At some point, most groups do agree on ways of working together and on acceptable behaviors, or norms, for the group. This important period in the group's formation defines whether, for example, the group will insist that all members attend meetings, whether insulting or other disrespectful remarks will be tolerated, and whether or not team members will be held to high or low standards for acceptable work. It is particularly important that team members understand and agree to the outcome of this so-called *norming* phase because it may well determine both the tone and the quality of subsequent work. Some characteristics of the norming phase include:

- clarification of roles in the group,
- emergence of informal leadership,
- development of a consensus on group behaviors and norms, and
- emergence of a consensus on the group's activities and purpose.

Significantly, norming is often the stage at which members decide just how seriously they are going to take the project. As such, it is important for team members who want a successful outcome to recognize that simply ignoring unacceptable behavior or poor work products will not be productive. For many teams, the norms of behavior that are established during the norming stage become the basis for behavior for the remainder of the project.

Performing: After the team has passed through the forming, storming, and norming stages, it should reach the stage of actively working on its project. This is the *performing* phase—the stage that most teams hope to reach. Here team members focus their energies on the tasks themselves, conduct themselves in accordance with the established norms of the group, and generate useful solutions to the problems they face. The characteristics of the performing phase include:

- clearly understood roles and tasks,
- well-defined norms that support the overall goals of the project,
- sufficient interest and energy to accomplish tasks, and
- emerging solutions and results.

This is the stage of team development in which is becomes possible for the goals of the team to be fully realized.

Adjourning: The last phase that teams typically pass through is referred to as *adjourning*. This stage is reached when the group has accomplished its tasks and is preparing to disband. Depending on the extent to which the group has forged its own identity, this stage may be marked by members feeling regret that they will no longer be working together. Some team members may act out some of these concerns in ways that are not consistent with the group's prior norms. These feelings of regret typically emerge after teams (or any groups) have been working together for a very long time, much longer than an academic semester or two because such complete group identity usually develops after a long time.

One final point about these stages of group formation. Teams will typically pass through each of them *at least once*. If the team undertakes significant changes in composition or structure, such as a change in membership, or a change in team leadership, it is likely that the team will revisit the storming and norming phases again.

2.4.1.2 *Team dynamics and brainstorming*

In Section 2.3.3 we discussed means of gaining information, generating and evaluating ideas, and obtaining feedback. Some of these means were based on putting the team and other stakeholders into situations that would encourage a free flow of ideas. Brainstorming and synectics in particular are based on the idea that one person's ideas may serve to stimulate other team members to come up with better alternatives. We now briefly summarize brainstorming and relate it to our discussion of the stages of group formation. We will see that our warnings to defer solutions until after the problem is sufficiently understood are not only consistent with our models of the design process, but also with our understanding of how teams best function.

Brainstorming is a classic technique for generating ideas and solutions to problems. Brainstorming consists of the members of a group offering individual ideas without any concurrent evaluation. Typically, a team will form a circle or sit around a table and, after a brief review of the problem for which ideas are being sought, offer ideas about the problem. One or more members of the team acts as the "scribe," writing down each idea offered for later discussion and review. Each member of the group should offer an idea when his or her turn comes, even if the idea is poorly formed or silly. At some point it becomes allowable for a team member to pass, but this must be done explicitly.

One of the anticipated outcomes of such brainstorming is an exhaustive list of potential solutions to the problem. Another hoped-for result is that one member's idea, even if impractical, may stimulate another member to leverage that first idea into a more useful one. It is extremely important that the participants *separate the generation of ideas from their evaluation.* Brainstorming is a technique for generating ideas that, by its nature, will lead to ideas that are subsequently rejected. At its core, brainstorming is an activity based on respect for the ideas of others, even to the point of being willing to suspend judgment on them temporarily. If a team focuses on evaluating ideas on the spot, it is likely to limit the willingness of team members to offer ideas. The team thus limits the extent to which creative changes or "piggybacked" ideas can emerge from earlier offerings.

> Separate the generation of ideas from their evaluation.

The previous discussion of group formation stages is relevant in helping to understand when a team can effectively engage in activities like brainstorming. Clearly in the forming and storming phases trust and confidence are likely to be absent from the team's dynamics. Indeed, the team may still be trying to define what the real task of the team is. The team is not likely to be in agreement about how seriously it will attempt to meet the team's nominal goals. As such, the team is almost certainly not going to be able to undertake effective brainstorming just yet. On the other hand, at the norming stage, the team is likely to be developing a consensus about norms of behavior, that may make it possible for the team to engage in the respect-based behavior that brainstorming requires. The team is most able to engage in brainstorming during the performing stage. This implies that models of design that allow for considerable early research and problem definition are most likely to be consistent with the underlying dynamics of how the team will perform.

2.4.2 Constructive conflict: Enjoying a good fight

Whenever people get together to accomplish tasks, conflict is an inevitable by-product. Much of this conflict is healthy, a necessary part of exchanging ideas, comparing alternatives, and resolving differences of opinion. Conflict can, however, be unpleasant and unhealthy to a group. It can also result in some team members feeling shut out or unwanted by the rest of the group. Thus, a solid understanding of the notions of constructive and destructive conflict is an essential starting point for team-based projects. Even in cases where team members have been exposed to conflict management skills and tools, it is useful to review them at the start of every project.

The notion of constructive conflict had its origins in research on management conducted in the 1920s. It was observed that the essential element underlying all conflict is a set of *differences*: differences of opinion, differences of interests, differences of underlying desires, etc. Conflict is unavoidable in interpersonal settings, so it should be understood and used to increase the effectiveness of all of the people involved. To be useful, however, conflict must be constructive. *Constructive conflict* is usually based in the realm of ideas or values. On the other hand, *destructive conflict* is usually based on the personalities of the people involved. If we were to list situations where conflict is useful or healthy, we might find such items as "generating new ideas" or "exposing alternative viewpoints." A similar list of situations in which conflict reduces a team's effectiveness would probably include items such as "hurting feelings" or "reducing respect for others."

> Constructive conflict is based on ideas and values.

The difference between *destructive, personality-based conflict* and *constructive, idea-based conflict* must be recognized by a team from the outset. While a team is establishing norms, and even before these have been formalized or agreed to in the "norming" phase, the team must establish some basic ground rules that prohibit destructive conflict, and it must enforce them by responding to violations of these ground rules. A team must not permit destructive conflict, including insults, personally denigrating remarks, and other such behaviors, from the outset, or it will become part of the team's culture.

Once we note this difference between constructive and destructive conflict, it is useful to recognize various ways that persons can react to conflict. Five basic strategies for resolving conflicts are:

- *avoidance*—ignoring the conflict and hoping it will go away;
- *smoothing*—allowing the desires of the other party to win out in order to avoid the conflict;
- *forcing*—imposing a solution on the other party;
- *compromise*—attempting to meet the other party "halfway"; and
- *constructive engagement*—determining the underlying desire of all the parties and then seeking ways to realize them.

The first three of these (avoidance, smoothing, and forcing), all turn on the notion of somehow making the conflict "go away." Avoidance rarely works, and serves to undercut the other party's respect for the person who is hiding from the conflict. Smoothing may be appropriate for matters where one or both of the parties in conflict really don't care about the issue at hand, but it will not work if the dispute is over serious, important matters. Once again, the respect of the person "giving in" may become lost over time. Forcing is only likely to be effective if the power relationships are clear, such as in a "boss-subordinate" situation, and even then, the effects on morale and future participation may be very negative. Compromise, which is a first choice for many people, is actually a very risky strategy for teams and groups. At its core, it assumes that the dispute is over the "amount" or "degree" of something, rather than on a true underlying principle or difference. While this may work in cases such as labor rates or times allocated for something, it is not likely to be effective in matters such as choosing between two competing design alternatives. (We cannot, for example, compromise between a tunnel and bridge by building a suspension tunnel.) Even in those cases where compromise is possible, we should expect that the conflict will be likely to reoccur after some period of time. Labor and management, for example, often compromise on wage rates—only to find themselves revisiting the very same ground as soon as the next contract opens up. That leaves us with constructive conflict as the only tool that holds the possibility of stable solutions to important conflicts.

Constructive conflict takes as its point of departure an honest telling of and listening to each party's underlying desire. Each side must reflect on what it truly wants from the conflict and honestly report that to the other parties. Each side must also listen carefully to what the other party really seeks. In many cases, the conflict is not based on the apparent problem, but rather because each party's underlying desires are different.

The originator of the idea of constructive conflict, Mary Parker Follett, told the following paradigmatic anecdote. She was working in a library at Harvard on a wintry day, with the windows closed. Another person came into the room and immediately opened one of the windows. This set the stage for a conflict and for identifying a way to resolve that conflict. All of the five resolution alternatives outlined above were available, but most of them were unacceptable. Doing nothing or smoothing would have left Ms. Follett uncomfortably cold. Compromising by opening the window halfway did not appear to be a viable alternative. Instead, she chose to speak with the other person and express her desire to keep the window closed in order to avoid the chill and draft. The other party agreed that this was a good thing, but noted that the room was very stuffy, which in turn bothered his sinuses. Both agreed to look for a reasonable solution to their underlying desires. They were fortunate to find that an adjacent work area also had windows that could be opened, thus allowing for fresh air to enter indirectly without creating a draft. Obviously, this solution was only possible because the configuration of the library allowed it. Nevertheless, they would not have even looked for this outcome except for their willingness to discuss their underlying desires. There are many cases where this will not work, such as when two people wish to marry the same third person. There are, however, many cases where constructive engagement will work, both to increase the solution space available to the parties in conflict and to heighten the understanding and respect of the other party. Even when the team is forced to revert to one of the "win-lose" strategies, the team should always first consider constructive engagement for resolving important conflicts.

2.5 CASE STUDY AND ILLUSTRATIVE EXAMPLES

John Maynard Keynes, a famous economist, once said that, "Nothing is required and nothing will avail, except a little, a very little, clear thinking." Design, however, is best learned by thinking and by doing. It is also fair to say that *design is best experienced by doing*. To that end, we strongly encourage fledgling designers and engineers to participate actively in teams doing design projects. Some of the formal techniques can be learned by doing exercises and observing how others have applied the techniques. To this end, we will elaborate a design case study and begin two design examples that illustrate the kinds of problems that engineers face when doing conceptual design, that is, when they have to devise concepts or ideas to solve a design problem.

2.5.1 Case study: Design of a microlaryngeal surgical stabilizer

We now present a case study of the conceptual design of a device that can help stabilize the instruments used during *laryngeal* or vocal cord surgery. This design project was undertaken by four teams of students in Harvey Mudd College's first-year design class on behalf of Brian Wong, M.D., Ph.D., of the Beckman Laser Institute of the University of California Irvine. The case study is an edited combination of results obtained by the four teams that shows *how a design team thought through the design process while they were designing a device for a client.*

Laryngeal, or vocal cord, surgery is often required to remove growths such as polyps or cancerous tumors. The "lead" cells of such growths must be removed accurately and completely. Patients also incur the risk of damage to their vocal cords—and so their speech—during these surgeries. In spite of many other surgical advances over recent decades, laryngeal surgery has not changed much. One change that has occurred is that surgeons now access the vocal cords through the mouth, rather than by cutting open the throat. This has made it harder to insert and stabilize optical devices and surgical instruments that cut, suck, grasp, move and suture. Surgeons must be able to control their own tremors in order make accurate and precise cuts during the procedure.

Tremor is the natural, small-scale shaking of the hand. (Watch the movement of your own fingertips as you hold your hands straight out in front of you.) In the context of laryngeal surgery, such tremors tend to be amplified as the surgeons insert and control foot-long instruments in the patient's throat.

The project began when Dr. Wong presented the following *initial problem statement* to the students who chose to work on this project:

> *Surgeons who perform vocal cord surgery currently use microlaryngeal instruments, which must be used at a distance of some 12–14 in. to operate on surfaces with very small structure (1–2 mm). The tremor in the surgeon's hand can become quite problematic at this scale. A mechanical system to stabilize the surgical instruments is required. The stabilization system must not compromise the visualization of the vocal cords.*

The four teams, each consisting of 3 or 4 students, talked with Dr. Wong and other physicians and did some basic library research to gain further information about laryngeal surgery. They learned that the abnormalities that were operated on were typically 1–2 mm wide, while the vocal cords themselves are approximately 0.15 mm wide. This meant that the physiological surgical tremors of the surgeon's hands had to be reduced from 0.5–3.0 mm to an acceptable tremor amplitude of 0.1 mm. They also learned that the surgeons needed to control the instruments at distances far from the patient's mouth (and vocal cords). One of the teams developed the following *revised problem statement*:

> *Microlaryngeal surgery seeks to correct abnormalities in the vocal cords. The abnormalities, such as tumors and cysts, are often 1–2 mm in size and are typically removed from the vocal cords, which are only 0.15 mm in size. During the operation, the surgeon must control his or her surgical instruments from a distance of 300–360 mm (12–14 in.) due to the difficulties in accessing the vocal cords. At this small scale, the physiological tremor in the surgeon's hand can be problematic. Design a solution that minimizes the effects of hand tremors in order to reduce unintentional movements at the distal end of the instrument to an amplitude of no more than 1/10 of a millimeter. The solution must not compromise visualization of the vocal cords.*

Note that this revised problem statement contains more detail and excludes an implied "mechanical" solution referenced in the original problem statement.

As part of this *information-gathering activity*, the teams worked to develop a list of the client's objectives for the designed stabilizing device, summarizing the attributes the client hoped the device would have and helping the design team to arrange those attributes in some priority order. The objectives and subobjectives are routinely displayed in an *objectives tree*. One team's objectives tree for this project is displayed in Figure 2.6. Two of the objectives are that the device should minimize obstruction of the surgeon's vision and that the cost of manufacture should be minimized. At the same time, the teams also developed lists of *constraints*, that is, the strict limits within which the designed device must remain. A constraint list for the device includes:

- it must be made of non-toxic materials;
- it must be made of materials that do not corrode;
- it must be sterilizable;
- its cost must not exceed $5,000;
- it must not have sharp edges;
- it must not pinch or gouge the patient; and
- it must be unbreakable during normal surgical procedures.

We see that there is an upper limit on the cost and on the device having sharp edges, among others.

The teams then set about choosing *metrics* that would enable them (at an appropriate future time in the design process) to determine whether various designs could actually achieve the objectives set out for the project. The metrics for two of the objectives of Figure 2.6, along with their units, are:

Objective: *Minimize viewing obstruction.*

Units: Rating percentage of view blocked on a scale of 1(worst) to 10 (best).

Metric: Measure the percentage of view blocked by the instrument. On a linear scale from 1 (100%) to 10 (0%), assign ratings to the percentage of view blocked.

Objective: *Minimize the cost.*

Units: Rating cost on a scale of 1(worst) to 5 (best).

Metric: Determine a bill of materials. Estimate labor, overhead, and indirect costs. Calculate the total cost. On a scale from 1(worst) to 5 (best), assign the following ratings to the calculated cost: $4,000–5,000 receives a 1; $3,000–4,000 receives a 2; $2,000–3,000 receives a 3; $1,000–2,000 receives a 4; and $1–1,000 receives a 5.

In addition to measuring the degree to which their objectives were achieved, the teams wanted to rank order their design objectives in terms of their perceived relative importance. This was done with an extension of what people normally do when comparing two objects, one against the other, called a *pairwise comparison chart* (PCC). The PCC allows each objective to be compared with every one of the other objectives. The PCC produced by one of the teams is shown in Table 2.1, and it shows that the most important objective is to reduce

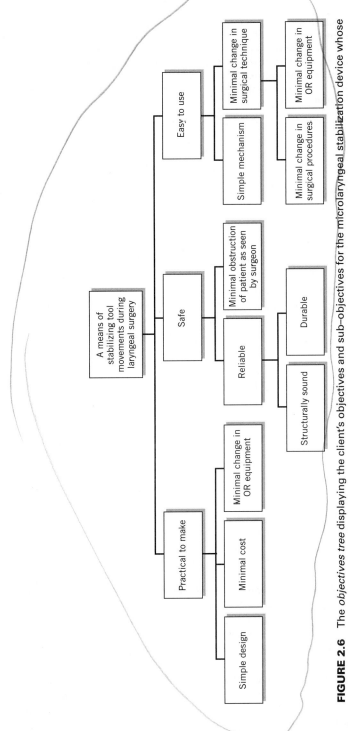

FIGURE 2.6 The *objectives tree* displaying the client's objectives and sub–objectives for the microlaryngeal stabilization device whose design is presented as a case study in Section 2.5.1. This tree is developed largely from the work of one of the three teams who worked on this project. Adapted from (Chan et al., 2000).

TABLE 2.1 A pairwise comparison chart created by one of the student teams to compare objectives for the microlaryngeal stabilization device. An entry "1" indicates the objective in that row is more important than that of the column in which it is entered. It shows that the reduction of the surgeon's tremor is the most important objective for this project. After (Both et al., 2000)

GOALS	Reduce tremor	Sturdy	Safe	Inexpensive	Easily used	SCORE
Reduce tremor	••••	1	1	1	1	4
Sturdy	0	••••	0	1	0	1
Safe	0	1	••••	1	1	3
Inexpensive	0	0	0	••••	0	0
Easily used	0	1	0	1	••••	2

the surgeon's tremor, while the least important is the cost of the instrument. This ranking helped focus the team's attention, as well as seeming to accord with our intuitions.

Having evolved to a deeper understanding of what the client wanted from this design, and of what is desired of the design's attributes, the design teams then turned to determining what a successful design will actually do. That is, the teams set about determining the *functions* that their proposed devices will perform, and writing the *performance specifications* that express in engineering terms how the performance of the functions can be measured and verified. The teams identified the required functions by applying some of the tools that will be discussed in Section 4.1, including the *black box*, the *glass box*, and the *functions-means tree*. One such list of functions states that the microlaryngeal stabilizer must:

- stabilize the instrument;
- move the instrument;
- stabilize the distal end of the instrument;
- reduce muscle tension of the surgeon during surgery (or reduce shaking tremors); and
- stabilize itself.

The performance specifications for the first of these functions was written as:

Function: *Stabilize the instrument.*

Performance specification: This function is not achieved if the design cannot reduce the amplitude of a trembling hand to less than 0.5 mm; it is optimally achieved if it controls the amplitude of a trembling hand to make it less than 0.05 mm; and it is overly restrictive if it inhibits or disallows any instrument or hand use.

With the functions and the specifications now largely determined, the design process turns to *creating alternative designs*. One excellent way to begin creating designs is to list each of the required functions in the left-hand column of a matrix, and list across each functional row the various *means* by which each function can be implemented. The resulting matrix or chart is called a *morphological chart* or "morph chart," and we show one for the microlaryngeal stabilizer in Figure 2.7. Such morph charts effectively tell us how large

FUNCTION	POSSIBLE MEANS					
Stabilize instrument	Hand	Stand	Clamp	Magnet	Edge of Laryngoscope	Wire	
Move instrument	Hand	Gears	Pneumatics	Ball bearing	Lever	Pulley	
Stabilize distal end of instrument	Magnet	Crosswires	Track system	Spring	Gyroscopes	Ball Bearings	Stand
Reduce muscle tension of surgeon during surgery	Instrument Stand	Hand Platform	Pillow	Elbow Platform	Forearm Rest	Shoulder Sling	
Stabilize itself	Gyroscope	Springs System	Stand	Magnet	Suspension System	Rest against stable surface	Attach to laryngo-scope

FIGURE 2.7   A *morphological chart* for the microlaryngeal stabilization device showing the *functions* and the corresponding *means* or *implementations* for each function. Possible designs are assembled in a "Chinese menu" fashion, that is, one from row *A*, one from row *B*, etc., etc., etc. Adapted from (Chan et al., 2000).

a *design space* we are working in because each design candidate must achieve every function, no matter which of its implementations or means are used. Thus, for each of the six (6) functions displayed in Figure 2.7, we combine (in "Chinese menu" fashion, with one from row *A*, one from row *B*, and so on) the following six (6) means to produce possible designs.

One candidate design has the instrument held in and moved by the surgeon's *hands*, with distal support provided by *crosswires*, which are attached directly to the laryngoscope. The surgeon uses a *forearm rest* to reduce tremor-inducing muscle tension.

A second design alternative has the instrument supported on a *stand*, moved by a system of *pulleys*, and supported by the *stand* itself. The *instrument stand* removes the need for surgeon to operate in a fixed position, which thus reduces tremor-inducing muscle tension.

As we will detail in our formal introduction of the morph chart, a given functional means does not necessarily connect with all of the means of all of the other functions, so there will inevitably be combinations that are excluded. Still, the combinatorial effects can be daunting for designs that have many functions, each of which can be implemented by several different means. Sooner or later, however, we will have to narrow the field of possible designs and, eventually, make a final choice on a final design. We do this by exercising

DESIGN CONSTRAINTS		LEVER	INSTRUMENT STAND	LARYNGOSCOPE CROSSWIRES
C: Must not fall apart during surgery		y	y	y
C: Noncorrosive materials		y	y	y
C: Must withstand medical sterilization procedures (autoclave, enzymatic, bleach, etc.)		y	y	y
C: Cannot get in the way of surgical instruments		y	y	y
C: Cannot block the view of vocal cords		y	y	y
C: Materials must be compatible with human body		y	y	y
C: Must be easy to clean with conventional means (scrub brush, water jet, soaking, etc.)		y	y	y
C: Cannot cost more than $5,000		y	y	y

DESIGN OBJECTIVES	Weight (%)	Score	Weighted Score	Score	Weighted Score	Score	Weighted Score
O: Structurally sound	15	0.75	11.25	0.85	12.75	0.80	12.00
O: Strong materials	11	0.85	9.35	0.9	9.9	0.85	9.35
O: Minimum obstruction of vocal cords	29	1	29	1	29	0.65	18.85
O: Minimum obstruction between patient and surgeon/nurse	9	0.65	5.85	0.7	6.3	1	9
O: Simple design	2	0.6	1.2	0.7	1.4	0.9	1.8
O: Minimum cost	2	0.5	1	0.7	1.4	0.55	1.1
O: Compatible with existing instruments	10	0.3	3	0.8	8	0.5	5
O: Minimal alteration of existing surgical procedures	8	0.45	3.6	0.8	6.4	0.7	5.6
O: Compatible with existing instruments	8	0.3	2.4	0.85	6.8	0.8	6.4
O: Simple mechanism	6	0.7	4.2	0.6	3.6	0.95	5.7
TOTAL	100		59.6		72.8		62.8

FIGURE 2.8 A *decision* or *selection matrix* used by one of the student teams that worked on the microlaryngeal stabilization device to select a final design. The decision matrix, whose numbers should be taken with caution, suggests which designs are preferred. Adapted from (Chan et al., 2000).

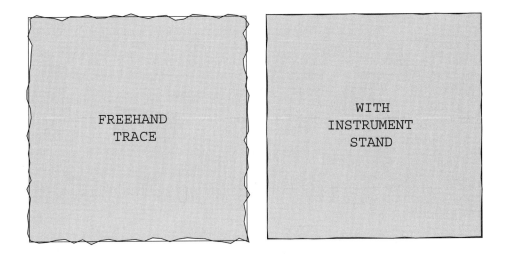

FIGURE 2.9 An example of the *testing* done by one student team to show that their concept successfully stabilized the surgeon's hand and reduced tremor, as demonstrated by the successful tracing of a predrawn square. Adapted from (Chan et al., 2000).

a *decision* or *selection matrix* in which we rate each possible design based on how well it achieves each of the design objectives and then add the scores for each to get a cumulative total for each design, as shown in Figure 2.8. As we will point out later, the results of this decision matrix should be used with care because many of the numbers are subjective (e.g., the weights assigned to each objective). In the present instance, for example, the design that was actually chosen by the client is the one that came in second in the selection process. That happened because the client realized only at the end—while trying out prototypes in a clinical setting—that he had not mentioned to the team an objective that turned out to be important!

Design concepts or ideas must also be tested in meaningful ways in order to ensure that they work. One of the student design teams tested their concept by attaching a pencil lead to the end of a surgical tool and tracing a pre-drawn square with the instrument, with and without their design attached. As we can see from their test results, displayed in Figure 2.9, their device removed almost all of the tremor.

Finally, after all of the work and the selections, the design process is complete, and a design is offered to the client. In the present case, one of the selected designs is being used in laryngeal surgeries and being prepared for manufacture as a medical product. Sketches of some of the team's final designs are shown in Figure 2.10, along with a drawing of the stabilizer being prototype-tested by Dr. Wong and his colleagues.

This case study has presented many of the design tools that are the main focus of this book. We have not shown the management tools, although the design teams on this project did use them, and we have said nothing about the dynamics of each of the three design teams. These as yet unsung elements are also very important for achieving effective design results.

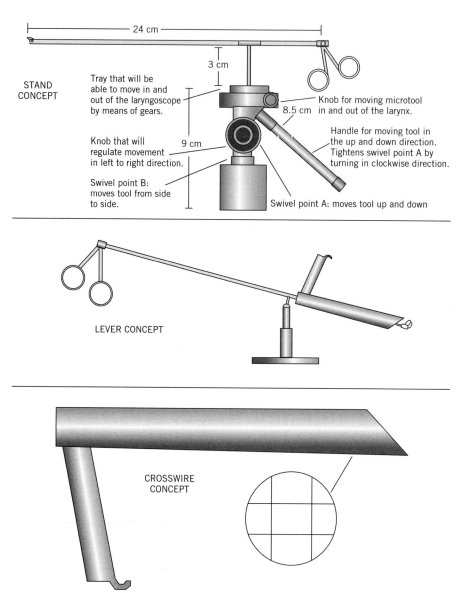

FIGURE 2.10 Several of the *design alternatives* produced by the student teams that worked on the microlaryngeal stabilization device for surgeons at the University of California at Irvine. Dr. Wong and his colleagues have adapted the crosswire concept for clinical trials. Adapted from (Chan et al., 2000, Saravonas et al., 2000).

2.5.2   Illustrative examples: Descriptions and project statements

We now describe the two illustrative examples that will be carried through the next four chapters to illustrate the various design techniques in developing yet familiar contexts. The first illustrative example is the design of a container for a new juice product,. It will be the design project for which we introduce and explain formal design methods. A summary of the beverage container project is:

> *Design of a container to deliver a children's beverage.* This is a stylized industrial design project that highlights some of the early questions that must be addressed before a designer can apply more conventional engineering science knowledge to the problem.
>
> Designers: Clive L. Dym and Patrick Little.
> Clients: Great American Food and Tobacco (GRAFT) and
>   Bringing Juice Into Children (BJIC).
> Users: Children living both in the United States and abroad.
> Project statement: Design a bottle for our new children's juice product.

The second illustrative design example is based on the work done by students in the first-year design course at Harvey Mudd College. We use their results, with both their permission and some post-project critiques of our own, to further explain how formal design methods are used. These further explanations appear at the ends of the chapters where particular formal design methods are introduced. This particular project is:

> *Design of a chicken coop to be built and used by a Mayan cooperative in Guatemala.* The chicken coop was designed by a team of Harvey Mudd College students in a freshman design class. It was subsequently built by the students on-site and of indigenous materials, with support from the humanitarian aid group Xela Aid.
>
> Designer(s): Teams of students in HMC's first-year design class.
> Client: XELA-Aid, a humanitarian aid group
> Users: Mayans living in the town of San Martin Chiquito, in the rural province
>   of Quetzaltenango, Guatemala
> Abbreviated project statement: Design and produce a chicken coop to be used
>   by several families in a Guatemalan village. The coop should be
>   made of indigenous materials and its design should take into
>   account the climate and environment of a Central American rain
>   forest.

2.6   NOTES

Section 2.1: The stepladder example derives from the freshman design course taught at Harvey Mudd College and is briefly described in (Dym, 1994b).

Section 2.2: As with definitions of design, there many descriptions of the design process and many of them can be found in (Cross, 1989), (Dym, 1994a), (French, 1985, 1992), (Pahl and Beitz, 1984), and (VDI, 1987). Further descriptions of the tasks of design can be found in (Asimow, 1962), (Dym and Levitt, 1991a), and (Jones,

1981). Examples of the application of conceptual design tools as problem-solving tools can be found in (Schroeder, 1998) for automobile evaluation and in (Kaminski, 1996) for college selection.

Section 2.3: Further elaboration of strategic thinking in design appears in (Dym and Levitt, 1991a). More detailed descriptions of the formal design methods can be found in (Cross, 1989), (Dym, 1994a), (French, 1985, 1992), (Pahl and Beitz, 1984), and (VDI, 1987). A strongly related discussion of concurrent engineering can be found in (Carlson-Skalak, Kemser, and Ter-Minassian, 1997). More detailed descriptions of the means for acquiring and processing knowledge are (also) given in (Bovee, Houston, and Thill, 1995), (Ulrich and Eppinger, 1995), and (Jones, 1992).

Section 2.4: The fundamentals of organizing teams are explained in (Tuckman, 1965) and cited in (Bartol, 1992). The discussion of Mary Parker Follett's explication of constructive conflict is adapted from (Graham, 1996). The basic stages of group formation are discussed in most recent management texts.

Section 2.5: The microlaryngeal stabilization device case study is detailed in (Both et al. 200, Chan et al., 2000, Feagan et al., 2000, and Saravanos et al., 2000). The two Xela-Aid chicken coop designs that we use as illustrative examples are derived from (Gutierrez et al., 1997) and (Connor et al., 1997).

2.7 EXERCISES

2.1 Describe in your own words the similarities and differences between the four models of the design process shown in Figures 2.1–2.4.

2.2 When would you be likely to use a descriptive model of the design process? When would you use a prescriptive model?

2.3 Map the management process shown in Figure 2.5 onto the design process shown in Figure 2.4.

2.4 Explain the differences between tasks, methods, and means.

2.5 You work for HMCI, a small engineering design company. You have been named team leader for a four-person design project that will be described in more detail in Exercises 3.2 and 3.5. You have not previously worked with any of these team members. Describe several strategies for moving the team quickly to the performing stage of group formation.

2.6 As Director of Engineering at HMCI, you notice that one of your team leaders and the team's client are unable to agree on a schedule. How might you advise the team leader to resolve this matter constructively?

CHAPTER 3

UNDERSTANDING THE CLIENT'S PROBLEM

What does this client want?

N THE preceding chapters we have defined engineering design, explored and described the process of design, and detailed some tools that we can use to monitor and control a design project. Now we turn to describing the tools we use in the preprocessing phase of design, during which we are working toward developing an engineering definition of the problem. The set of activities done during the preprocessing phase is also known as *problem definition*.

3.1 OBJECTIVES TREES: TRANSLATING AND CLARIFYING THE CLIENT'S WANTS

The starting point of most design projects is the identification by a *client* of a *need* to be met. The fulfillment of that need then becomes the goal of the chosen design team. As depicted in the models of the design process in Figures 2.1–4, the client's need is quite often presented as a verbal statement in which the client identifies a gadget that will appeal to certain markets (e.g., a container for a new beverage), a widget that will perform some specific functions (e.g., a chicken coop), or a problem to be fixed through a new design (e.g., a new transportation network and hub).

Sometimes clients' project statements are quite brief. For example, consider the beverage container problem described in Section 2.5. Whether the design team is working for Great American Food and Tobacco (GRAFT) or for Bringing Juice Into Children (BJIC), it might simply get a memo from upper management that says: "Design a bottle for our new children's fruit juice product." The design team could easily respond to this directive by choosing an existing bottle, slapping on a clever label, and calling its work done. However, we might ask whether this new bottle is a *good* design, or, further, whether it's the *right* design. Answers to these questions depend on how we measure goodness and on how we assess rightness or correctness. Thus, we will also discuss metrics against which we can measure designs in this chapter.

Another simple project statement might take the form of "The Claremont Colleges need to reconfigure the intersection of Foothill Avenue and Dartmouth Avenue so students can cross the road." While communicating someone's idea of what the problem is, statements like this one have limitations because they often contain errors, show biases, or imply solutions. *Errors* may include incorrect information, faulty or incomplete data, or simple mistakes regarding the nature of the problem. Thus, the problem statement just given should refer to Foothill Boulevard, not Foothill Avenue. *Biases* are presumptions about the situation that may also prove inaccurate because the client or the users may not fully grasp the entire situation. In the traffic example, for instance, the real problem may not be related to the design of the intersection but to the timing of the signal lights or to the tendency of students to jaywalk. *Implied solutions*, that is, a client's best guesses at solutions, frequently appear in problem statements. While implied solutions offer some useful insight into what the client is thinking, they may wind up restricting the design space in which the engineer searches for a solution. Also, sometimes the implied solution fails to actually solve the problem at hand. For example, it is not obvious that reconfiguring the intersection will solve the student traffic problem. If students jaywalk, reconfiguring the intersection will do little or nothing to mitigate this. If the problem is that students are crossing a dangerous street, we may want to relocate the destination to which they are headed. The point is that we must

> The client's understanding of the problem usually requires clarification by the designer.

carefully examine project statements in order to identify and deal with errors, biases, and implied solutions. Only then do we get to the real problem.

For now, we want to focus on developing a clearer understanding of what the client wants because this will help us see the lines along which measures for a design might emerge. That is, we want to clarify what the client wants, account for what potential users need, and understand the technological, marketing, and other contexts within which our gadget or widget will function. In so doing, we will be *defining our design problem* much more clearly and realistically. (We will see that this is where we start to think about what will emerge as the *product specifications*, the formal statements of the properties and functionalities that our design must have.)

3.1.1   Object attributes and lists of objectives

Imagine that we are members of a design team that is consulting for a company that makes both low- and high-precision tools (with corresponding prices). The company's management, seeking to penetrate a new market, has given the team a charter more specific than designing a "safe ladder," to wit, "Design a new ladder for electricians or other maintenance and construction professionals working on conventional job sites." This is a fairly "routine" design task, but to really understand the goals of this design, we still need to talk with management, some potential users, some of the company's marketing people, and some experts. We also need to conduct our own brainstorming sessions. We will get a better understanding of what our design project is really about by asking questions such as:

- What features or attributes would you like the ladder to have?
- What do you want this ladder to do?
- Are there already ladders on the market that have similar features?

And while asking these three questions, we will also want to ask:

- What does that mean?
- How are you going to do that?
- Why do you want that?

As a result of our discussions and brainstorming, we might generate the list of characteristics and attributes of a safe ladder design shown in List 3.1

List 3.1   *SAFE LADDER Attributes List*

Ladder should be useful

Used to string conduit and wire in ceilings

Used to maintain and repair outlets in high places

Used to replace light bulbs and fixtures

Used outdoors on level ground

Used suspended from something in some cases

Used indoors on floors or other smooth surfaces

Could be a stepladder or short extension ladder

A folding ladder might work

A rope ladder would work, but not all the time

Should be reasonably stiff and comfortable for users

Step deflections should be less than 0.05 in

Should allow person of medium height to reach/work at levels up to 11 ft

Must support weight of an average worker

Must be safe

Must meet OSHA requirements

Must not conduct electricity

Could be made of wood or fiberglass, but not aluminum

Should be relatively inexpensive

Must be portable between job sites

Should be light

Must be durable

Needn't be attractive or stylish

We note in examining this list that not all of the statements are of the same kind. Some of them can be considered as binary issues, answered either by a "Yes" or a "No," while others allow for a range of answers. For example, the statement "Must not conduct electricity" doesn't really leave any options: either the ladder is a conductor or it is not a conductor, and in this case it must be an insulator. On the other hand, statements such as "Should be relatively inexpensive" allow for a range of prices at which the ladder could be built and sold. A ladder that can be built for $15 is more desirable than one that can be built for $20, assuming all the other characteristics are the same.

There are also other differences in these statements. The material of which the ladder is to be made (wood or fiberglass, but not aluminum) is a design choice that should probably be deferred until later in the design process, unless there is some specific information from the client that forces an early choice. The idea that the ladder must resist certain forces in certain ways (e.g., a limit on the deflection of a step) is a reflection of the way that engineers begin to translate features of a design into specifications (a subject we return to in

Chapter 5). The differences among the statements in List 3.1 are of considerable interest to the designer as they reflect differences between fundamentally different intellectual objects, a subject we address in Section 3.2.1.

3.1.2   Goals and objectives, constraints, functions, and implementations

The reason that the statements in the foregoing list for the safe ladder seem different in kind is that they reflect different intellectual objects that must be considered and evaluated by a designer. There are clearly some *objectives* we want to achieve (e.g., the ladder should be relatively inexpensive), some *constraints* (e.g., a step should deflect no more than 0.05 in.), some *functions* (e.g., to support workers), and there are some *means* or *implementations* (e.g., the ladder could be made of wood or fiberglass). We have already defined some of these terms in Section 1.2, but we will review them here with the intent of being able to recognize them in the context of an attribute list having different kinds of statements that would emerge very early in a design project, such as the safe ladder list just given.

Objectives or *goals* are ends that the design strives to achieve. (We generally view design objectives and design goals as meaning much the same thing, with the possible exception that reference to *the* design goal as the top-level objective.) Objectives are expressions of the desired attributes and behavior that the client or potential users would like to see in the designed object. They are normally expressed as "being" statements that say what the design will *be*, as opposed to what the design must *do*. For example, saying that the ladder should be portable is a "being" term. "Being" terms identify attributes that make the object "look good" in the eyes of the client or user, expressed in the natural languages of the client and of potential users.

> Objectives are the desired attributes of a design.

Another way of identifying objectives is to note that they are often written as statements that "more (or less) of [the objective]" is better than "less (or more) of [the objective]." For example, lighter is better than heavier if our goal is portability. As such, objectives lend themselves to being measured somehow. In this way, objectives help us to choose among alternative design configurations.

Constraints are restrictions or limitations on a behavior or a value or some other aspect of a designed object's performance. Constraints are typically stated as clearly defined limits whose satisfaction can be framed into a binary choice, for example, the ladder material is a conductor or it is not, or the step deflection is less than 0.05 in. or it is not. Constraints are important to the designer because they limit the size of a design space by forcing the exclusion of unacceptable alternatives. For example, any ladder design that fails to meet OSHA standards will be rejected.

> Constraints are (strict) limits that a design must meet to be acceptable.

Objectives and constraints sometimes seem to be interchangeable, but they are not. They are, however, closely related. Constraints limit the size of the design space, while objectives permit the exploration of the remainder of the design space. That is, constraints can be formulated to allow us to reject alternatives that are unacceptable, while objectives allow us to select among design alternatives that are at least acceptable, or, in other words, they *satisfice*. Designs that satisfice (or alternative selections in situations that require a choice to be made) may not be the best or optimal, but they do, at least, minimally satisfy all constraints. For example, we could barely satisfy OSHA standards or we could satisfy them "in spades" by making a "super safe" ladder in order to obtain a marketing advantage. Or, on the price side, a goal that a ladder should be "low cost" could be cast as a statement that the cost to build the ladder cannot exceed $25. In that case we have a fixed limit, a constraint. If we have *both* a low-cost objective and a $25 constraint, we may be able to exclude some initial designs based on the constraint alone, leaving us free to choose among the remaining designs based on cost and other, noneconomic objectives.

It is important to recall that both objectives and constraints *refer to the object being designed*, not to the design process. The "low-cost ladder," for example, has a low manufacturing or production cost. The cost of the design process (engineering salaries, market surveys, prototype development, etc.) may be high, but that's a separate matter altogether.

Functions are the things a design is supposed to *do*, the actions that it must perform. In our initial attributes list, functions are usually expressed as "doing" terms that often reflect the language of the engineer. We will discuss functions in greater detail in Chapter 4.

Lastly, *implementations* or *means* are ways of executing those functions that the design must perform. These are the items on the attributes list that provide specific suggestions about what a final design will look like or be made of (e.g., the ladder will be made of wood or of fiberglass), so they often appear as "being" terms. However, it is usually pretty obvious which "being" terms are goals to be achieved and which "being" terms are very specific properties. It is premature for us to consider means in great detail here, since any means or implementations that we might select would be governed by the things that a specific designed object must do. That is, implementations and means are very much *solution-dependent* in that they are often design choices made to implement the functions that are to be performed by the already-chosen design.

We can now pare the attributes List 3.1 by removing or pruning the constraints, functions, and implementations, leaving only objectives on the list. Thus, our pruned list of objectives for the ladder is given in List 3.2.

> **List 3.2   SAFE LADDER Pruned Objectives List**
>
> Ladder should be useful
>
> Used to string conduit and wire in ceilings
>
> Used to maintain and repair outlets in high places
>
> Used to replace light bulbs and fixtures
>
> Used outdoors on level ground
>
> Used suspended from something in some cases
>
> Used indoors on floors or other smooth surfaces
>
> Should be reasonably stiff and comfortable for users
>
> Should allow person of medium height to reach/work at levels up to 11 ft
>
> Must be safe
>
> Should be relatively inexpensive
>
> Must be portable between job sites
>
> Should be light
>
> Must be durable

(margin note) Functions are actions that a successful design must perform.

(margin note) Implementations are specific choices of design options.

While List 3.2 is useful as a list of goals to be achieved, there is much more that we can do with it. In particular, if our list was much longer, we might find it difficult to use the list without organizing it in some way. We may want to group or *cluster* these objectives together in some coherent way. One way to start grouping entries on the list is to ask ourselves why we care about them. For example, why do we want our ladder to be used outdoors? The answer is probably because that's part of what makes a ladder useful, which is another entry on our list. Similarly, we could ask why we care whether the ladder is useful. In this case, the answer is not on the list—we want it to be useful so that people will buy it. Put another way, usefulness makes a ladder marketable. This suggests that we need an item on our list about marketing, for example, "The ladder should be marketable." This turns out to be a very helpful objective, since it tells us why we want the ladder to be cheap, portable, etc. If we go through clustering questioning of this sort, we will find a new list that we can put in the form of an *indented outline*, with *hierarchies* of major headings and various degrees of subheadings, as shown in List 3.3.

List 3.3 *SAFE LADDER Indented Objectives List*

0. *A safe ladder for electricians*

  1. The ladder should be safe

      1.1 The ladder should be stable

          1.1.1 Stable on floors and smooth surfaces

          1.1.2 Stable on relatively level ground

      1.2 The ladder should be reasonably stiff

  2. The ladder should be marketable

      2.1 The ladder should be useful

          2.2.1 The ladder should be useful indoors

              2.2.1.1 Useful to do electrical work

              2.2.1.2 Useful to do maintenance work

          2.2.2 The ladder should be useful outdoors

          2.2.3 The ladder should be of the right height

      2.2 The ladder should be relatively inexpensive

      2.3 The ladder should be portable

          2.3.1 The ladder should be light in weight

          2.3.2 The ladder should be small when ready for transport

      2.4 The ladder should be durable

As we can see, this revised, indented outline allows us to explore each of the top-level objectives further, in terms of the subobjectives that tell us how to realize it. At the highest level, our objectives turn us back to the original design statement we were given, namely to design a safe ladder that can be marketed to a particular group.

Now, we have certainly not exhausted all the questions we could ask about the ladder, but we can identify in this outline some of the answers to the three questions mentioned just above. For example, "What do you mean by safe?" is answered by two subgoals in the cluster of safety issues, that is, that the designed ladder should be both stable and relatively stiff.

We have answered "How are you going to do that?" by identifying several subgoals or ways in which the ladder could be useful within the "The ladder should be useful" cluster and by specifying two further "sub-subgoals" about how the ladder would be useful indoors. And we have answered the question "Why do you want that?" by indicating that the ladder ought to be cheap and portable in order to reach its intended market of electricians and construction and maintenance specialists.

We can represent the indented outline we have just started in graphical form simply by laying out a *hierarchy* of boxes, each of which contains an objective for the object being designed, as shown in Figure 3.1. Each layer or row of objective boxes corresponds to a level of indentation (which is indicated by the number of digits to the right of the first decimal point) in the outline. Thus, the indented outline becomes an *objectives tree*: A graphical depiction of the objectives or *goals for the artifact* (as opposed to goals for a design project or process). The top-level goal in an objectives tree, which we represent as a node at the peak of the tree, is decomposed or broken down into subgoals that are at differing levels of importance or that include progressively more detail, so the tree reflects an *hierarchical structure* as it expands downward. An objectives tree also shows that related subgoals or similar ideas can be *clustered* together, which gives the tree some organizational strength and utility.

Still further, the graphical format of the tree is quite useful for discussions with clients and other participants in the design process. It is also useful for determining what things we need to measure, since we will use these objectives to decide among alternatives. The graphical format or tree is also useful since it corresponds to the mechanics of the process that many designers follow. Often, the most useful way of "getting your mind around" a large list of objectives is to put them all on Post-It™ notes, and then move them around until the design team is satisfied with the tree. We will discuss some of the mechanics of tree building and problem definition in Section 3.5.

The process just outlined—from lists to refined lists to outlines to trees—has a lot in common with one of the fundamental skills of writing, being able to construct an outline. A

> Objectives trees are ordered lists of the desired attributes of a design.

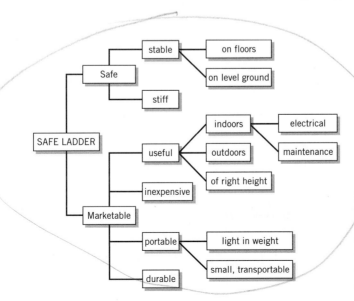

FIGURE 3.1 The objectives tree for the design of a safe ladder. It shows the first fruits of problem definition. Note the hierarchical structure and the clustering of similar ideas.

topical outline provides an indented list of topics to be covered, together with the details of the subtopics corresponding to each topic. Since each topic represents a goal for the material to be covered, the identification of an objectives tree with a topical (or an indented) outline seems logical.

We answer "how" by digging deeper into an objectives tree.

One final point about this simple example. Note that as we work *down* the tree, or move further in on the levels of indentation, we are doing more than just getting into more detail. We are also answering a generic *how* question for many aspects of the design, that is, the question of "*How* are you going to do that?"

We answer "why" by searching higher on an objectives tree.

Conversely, as we move *up* the tree, or further out toward fewer indentations, we are answering a generic *why* question about a specific (needed) function, that is, the question of "*Why* do you want that?" This enables us to track why we want some feature or other fine point in our design, which may be very important if we have to trade off features, one against the other, because the values of these features may be directly attributable to the importance of the goals they are intended to serve. We will say more about this in Section 3.3.

3.1.3   How deep is an objectives tree? What about pruned entries?

Where do we end our list or tree of objectives? The simple answer is to stop when we run out of objectives or goals and implementations begin to appear. That is, within any given cluster, we could continue to parse or decompose our subgoals until we are unable to express succeeding levels as further subgoals. The argument for this approach is that it points the objectives tree toward a *solution-independent* statement of the design problem. That is, we know what characteristics the design has to exhibit without having to make any judgment about how it might get to be that way. In other words, we determine the attributes of the designed object without specifying the way the objective is realized in concrete form.

Another way of limiting the depth of an objectives tree is to look out for verbs or "doing" words because they normally suggest functions. Functions should definitely not appear on objectives trees or lists.

A second tree-building issue is deciding what to do with the things that we have removed from the list. In the case of the functions and implementation, we simply put them aside (recording them in case they are good ideas), and pick them up again later in the process. In the case of constraints, however, it is often reasonable to reenter them into an appropriate place in the objectives tree, while being careful to distinguish them from the objectives. For example, in an outline form of the objectives tree, we might use italics or a different font to denote constraints (see List 3.4 in the next section). In a graphical form, we may wish to highlight constraints using differently shaped boxes. In either case, it is important to recognize that constraints are related to but different from objectives, and they are used in different ways.

3.1.4   The objectives tree for the beverage container design

In the beverage container design problem, our design team is working for one of the two competing food product manufacturers, in this instance BJIC. (We note parenthetically that there is an interesting ethical problem that we will address in Chapter 10, that is, could our design team, or our firm, take on the same or similar design tasks for both, or for two competing clients?) However, for now, let us suppose that we're dealing with a single client and that our client's project statement is as stated in Section 3.5: "Design a bottle for our new juice product."

In order to clarify what was wanted from this design, our design team questioned many people in BJIC, including the marketing staff, and we talked to some of their potential customers or users. As a result, we found that there were several motivations driving the desire for a new "juice bottle," including: plastic bottles and containers all look alike; the client, as a national producer, has to deliver the product to diverse climates and environments; safety is a big issue for parents whose children might drink the juice; many customers, but especially parents, are concerned about environmental issues; the market is very competitive; parents (and teachers) want children to be able to get their own drinks; and, finally, children always spill drinks.

These motivations emerged during the questioning process, and their effects are displayed in the augmented attributes list for the container given as List 3.4. Some of the entries in this list are shown in italics because they are constraints. Thus, these constraint entries can be removed from a final list of the attributes that are objectives (to be reinserted later, as discussed above).

List 3.4   *BEVERAGE CONTAINER Augmented Attributes List*

Safe	⟶	DIRECTLY IMPORTANT
Perceived as Safe	⟶	Appeals to Parents
Inexpensive to Produce	⟶	Permits Marketing Flexibility
Permits Marketing Flexibility	⟶	Promotes Sales
Chemically Inert	⟶	*Constraint* on Safe
Distinctive Appearance	⟶	Generates Brand Identity
Environmentally Benign	⟶	Safe
Environmentally Benign	⟶	Appeals to Parents
Preserves Taste	⟶	Promotes Sales
Easy for Kids to Use	⟶	Appeals to Parents
Resists Range of Temperatures	⟶	Durable for Shipment
Resists Forces and Shocks	⟶	Durable for Shipment
Easy to Distribute	⟶	Promotes Sales
Durable for Shipment	⟶	Easy to Distribute
Easy to Open	⟶	Easy for Kids to Use
Hard to Spill	⟶	Easy for Kids to Use
Appeals to Parents	⟶	Promotes Sales
Chemically Inert	⟶	*Constraint* on Preserves Taste
No Sharp Edges	⟶	*Constraint* on Safe
Generates Brand Identity	⟶	Promotes Sales
Promote Sales	⟶	DIRECTLY IMPORTANT

The augmented List 3.4 also shows how, after additional brainstorming and questioning, some of the listed goals are either expanded into subobjectives (or subgoals) and others are connected to existing goals at higher levels. In one case a brand new top-level goal, Promote Sales, is identified. The objectives tree corresponding to (and expanded from) this augmented attribute list is shown in Figure 3.2, and a tree combining objectives and constraints is shown in Figure 3.3. The detailed subgoals that emerge in these trees clearly track well with the concerns and motivations identified in the clarification process.

As a result of the thought and effort that went into List 3.4 and the objectives trees of Figures 3.2 and 3.3, the design team rewrote and revised the problem statement for this design project to read: "Design a safe method of packaging and distributing our new children's juice product that preserves the taste and establishes brand identity to promote sales to middle-income parents." Thus, as we noted in Chapter 2, one of the outputs of the design preprocessing (or problem definition) phase is a revised statement that reflects what has been learned about the goals for a design project. That is, the emergence of a clearer understanding of the client's design problem results in an objectives tree that points toward the expression of the features and behaviors wanted from the designed object, and it often results in a simultaneous revision or restatement of the client's original problem statement.

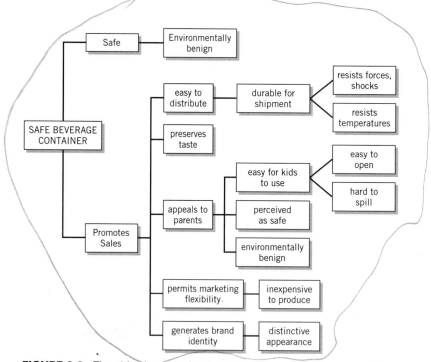

FIGURE 3.2 The objectives tree for the design of a new beverage container. Here the work on problem definition has lead to a hierarchical structuring of the needs identified by the beverage company and by the potential consumers—or at least the consumers' parents!—of the new children's juice drink.

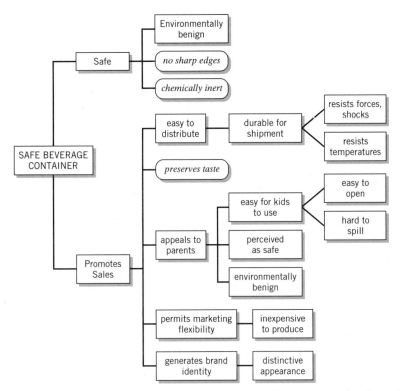

FIGURE 3.3 A combined tree (objectives in rectangles and constraints in ovals) for the design of a new beverage container. Here the goals for the new product are shown together with the constraints that apply to the object being designed.

3.2 CONSTRAINTS: SETTING LIMITS ON WHAT THE CLIENT CAN HAVE

There are limits to everything. That is why constraints are extremely important in engineering design, as we noted in Section 3.1.2 when we articulated some differences between constraints and objectives.

Constraints enable us to identify and exclude unacceptable designs.

As a practical matter, many designers use constraints as a sort of "checklist" for designs in order to prune the set of designs to a more manageable size. Such constraints, which can be included in properly marked trees that include both objectives and constraints, are usually expressed in terms of specific numbers. By way of contrast, objectives are normally expressed as verbal statements that sometimes can be formulated in terms of continuous variables or numbers that may allow for a range of values of interest to the designer. To reiterate our earlier illustration of this point, a goal that a ladder should be cheap could be stated in terms of having a materials or manufacturing cost that does not exceed a fixed limit or constraint, say $25. On the other hand, we could have *both* an objective that the ladder be cheap *and* a constraint that puts a limit on the cost. In that case we are able to choose among a set of designs whose costs to build are different, as long as they are all below the limit set by the constraint alone. This, again, is the strategy of "satisficing" wherein we select among design alternatives that are acceptable.

There is still another approach to dealing with objectives that can be cast in terms of "continuous variables." There are many design domains in which we can formulate mathematical relationships between many of the design variables. For example, we might know how the cost of the ladder depends on its weight, its height, the size of its projected market, and other variables. In such cases, we may try to *optimize* or get the best design, say the minimum cost ladder, using procedures much the same as we use to find the maxima or minima of multi-variable calculus problems. Similarly, *operations research* techniques allow calculations to be performed when the design variables are explicitly numerical. Optimization techniques are clearly beyond the scope of our discussions, but the underlying idea that design variables, and design objectives, interact and vary with one another, is also a theme that we will further elaborate in the following section when we discuss ways of assessing the comparative values of design goals.

3.3 SETTING PRIORITIES: RANK ORDERING WHAT THE CLIENT WANTS

We have been rather insistent in this chapter that we properly identify and list all of the client's objectives, taking care also to ensure that we don't mix up constraints, functions, or means with the goals set for the object being designed. But do we know that all of the identified objectives have the same import or value to the client or the users? So far we have assumed that each of the top-level objectives has the same value to all concerned because we have made no effort to see whether there is any variation in perceived value. It is almost certain that some goals will be more important than others, so we ought to be able to recognize that and measure it. How are we going to do that?

3.3.1 On measuring things

Let us step back for a moment and think about what it means to measure and compare some design objects, whether they be the objectives for a design, a set of design attributes, or a collection of conceptual designs. It is not obvious that there is any way to meaningfully mark or "plot" the individual design objects along an axis, but it does makes sense to plot evaluation points earned by each object along a line. Those evaluation points can be compared and some design decisions can be taken. How do we award such evaluation points? Is there a scale we can use to lay out the evaluation points for each design?

Engineers are used to measuring all sorts of things: beam lengths, surface areas, hole diameters, speeds, temperatures, pressures, and so on. In each of these cases, there is a ruler involved that shows a zero and has marks that show units, whether they be inches, microns, mm of Mercury, or degrees Farenheit or Centigrade. This establishes a common basis for comparing. Without rulers, how would we meaningfully quantify the assertion that "*A* is taller than *B*"? Simply standing *A* and *B* against each other, back to back, doesn't cut it (especially if *A* and *B* are not easily moved). However, by using a measuring stick that has a zero and is marked with fixed intervals of length that can be counted, we can establish real numbers to represent the heights of *A* and *B*.

The important concept here is that of having a *ruler* with (1) a *defined zero*, and (2) a *unit* that is used to define the markings scribed onto the ruler. In mathematical terms, these properties enable *strong measurement*, as a consequence of which we can treat measured mathematical variables (say *L* for length, *T* for temperature, and so on) as we would any variable in calculus. Thus, strong measurements could be used as would any of our "normal" physical variables in a mathematical model.

Six different types of scales have been used to evaluate and test product designs, as shown in Table 3.1. These different types of "scales" and their associated units of measure can each be used in different situations, but *there are limits to what can be done with these "measurements"* because some of them are not "real" measurements. *Nominal scales*, for example, are used to distinguish between categories. We can count the number of colors available, but there is no measure of color difference. The same can be said of *partially ordered scales*, such as hierarchies of families. Thus, these scales are of little use for examining most design choices, even if the distinctions drawn are of interest to the client, the users, or the designer.

Ordinal scales are used to place things in rank order, that is, in first-, second-, or *n*th place. This seems straightforward enough, but it is precisely here that measurement gets more complicated when it comes to assessing the *subjective preferences* of individuals. That is, when we ask the client which design objectives are most important, we are typically asking for a subjective ranking of their perceived importance. To ask whether cost or portability is the more important objective in the design of a ladder, as we will discuss in Section 3.3.2, is to ask a question to which the answer is somehow different than the statement that "August Dym Noë is 2 ft., 1 in. tall." We can indicate a preference for portability over price, but there is no sensible way to assess the degree or amount of that preference. For example, there is no meaningful way to say that "portability is five times more important than cost" because there is no scale or ruler that defines both a zero and a unit with which to make such measurements.

Ratio scales have naturally defined base points that have physical meaning (i.e., zero money, of zero height, etc.) and can be measured. In the case of objectives, ratio scales for

Table 3.1 Measuring scales for testing and evaluating designs in the field of product design. Adapted from (Jones, 1992).

Nominal scales, such as colors, smells, or even professions (e.g., teachers, lawyers, engineers).

Partially ordered scales, such as grandparent, parent, and child, which array themselves somewhat in order of seniority.

Ordinal scales, such as first, second, third, etc.

Ratio scales, such as inches, seconds, or dollars. Ratio scales have natural reference points or base points.

Interval scales, such as degrees Centigrade, that have arbitrarily defined reference points or base points.

Multidimensional scales or *index numbers*, such as miles per gallon or kilometers per maintenance event, that are compounds of other scales of measurement.

design objectives would have specific values that can be understood as "zero." For example, the notion that a product will emit no pollution is straightforward.

Interval scales have defined reference points or base points, from which all others are referenced (or, to which all others are related). Interval scales are the closest of the traditional scales to what we have earlier identified as strong measurement.

Assessing objectives often involves measurements for which there is no ruler. If "simplicity" of a product is a design objective, how is simplicity to be measured? The answer is that a *metric* would be introduced, for example, the number of parts. A certain minimum number of parts would be identified as a base point, and other designs could be assessed by the number of parts they contain as compared to the base number of parts. Metrics will be discussed further in Section 3.4.

Given the disparate nature of design objectives or of a set of designs for a product, it is far from clear that an identifiable scale or ruler can be meaningfully used to assess and evaluate either objectives or designs. It would be easier, for example, to evaluate designs by their estimated manufacturing costs, which are hard numbers measurable on a standard ratio scale. However, for both proposed designs and for sets of design objectives, we are trying to assess subjective preferences that are not easily rendered quantitative.

3.3.2 Pairwise comparison charts: One way to rank order things

Suppose we have a set of goals for a project whose relative values we want to *rank;* that is, we want to identify their value or importance relative to one another and to order them accordingly. Sometimes we get really lucky and our client expresses strong and clear preferences, or perhaps the potential users do, so that the designer doesn't have to do an explicit ranking. More often, however, we do have to do some ranking or we have to place some values ourselves. Thus, we propose here a fairly straightforward technique that can be used to rank goals that are at the same level in the hierarchy of objectives and are within the same grouping or cluster; that is, they have the same parent or antecedent goal within the objectives tree. It is very important that we make our comparisons of goals with these clustering and hierarchical restrictions firmly in mind in order to ensure that we're comparing apples with apples and oranges with oranges. For example, does it make sense to compare the subgoals of having a ladder be useful for electrical work and having it be durable? On the other hand, rank ordering the importance of the ladder's usefulness, cost, portability, and durability would be good design information.

Suppose we are designing just such a ladder for which four high-level goals have been established: It should be inexpensive, useful, portable, and durable. Let us further suppose that we can easily choose between any given pair of them. For example, we might prefer cost over durability, portability over cost, portability over usefulness, and so on. Where does this leave us in terms of ranking all four goals? We can determine one answer to that question by constructing a simple chart or matrix that allows us to (1) compare every goal with each of the remaining goals individually, and (2) add cumulative or total scores for each one of the goals.

We show in Table 3.2 a *pairwise comparison chart* (PCC) for our four-objective ladder design. The entries in each box of the chart are determined as binary choices, that is, every entry is either a 1 or a 0. Along the row of any given goal, say Cost, we enter a zero in

Table 3.2 A pairwise comparison chart (PCC) for a ladder design.

Goals	Cost	Portability	Usefulness	Durability	Score
Cost	• • • •	0	0	1	1
Portability	1	• • • •	1	1	3
Usefulness	1	0	• • • •	1	2
Durability	0	0	0	• • • •	0

Pairwise comparisons help us understand the relative importance of the items (e.g., objectives) being compared.

those columns for the goals Portability and Usefulness that are preferred over Cost, and we enter 1 in the Durability column because Cost is preferred over Durability. We also enter zeroes in the diagonal boxes corresponding to weighting any goal against itself, and we enter ratings of 1/2 for goals that are equally valued. The scores for each goal are determined simply by adding across each row. We see that in this case, the four goals can be ranked (with their scores) in order of decreasing value or importance, Portability (3rd), Usefulness (2nd), Cost (1st), and Durability (4th).

We might wonder whether a score of 0 means that we can or should drop durability as an objective, but it takes only a little bit of clear thinking to see that we *cannot* drop objectives that score zeroes. The zero means only that of the four ranked objectives, durability was ranked fourth and least important. And, since the ranking numbers only indicate place in line, they will not be used in subsequent calculations and the zero cannot get us into trouble.

Now, the simple PCC process just described is a valid way of ordering things, but its results should be taken as *no more than a straightforward rank ordering*, or an ordering of place in line. The scores assembled in Table 3.2 do not constitute what we had defined as (mathematically) strong measurement because there is no rational scale on which to measure these very different objectives, and there is only an implied, but undefined zero. Thus, these scores should not be used as entries in further calculations. They are a useful guide for further thought and discussion, but they are *not* a basis for further calculations.

3.3.3 Pairwise comparison charts: Consensus rank orderings

Life is still more complicated when assessing the preferences of groups. We have been working in the framework of the single designer or decision maker who is making a subjective assessment, determined to obtain a meaningful and useful ranking. The group situation—in which members of a design team vote on their preferences so that their individual votes can be gathered into an aggregated, group set of preferences—is even more complicated and a subject of much debate. (See the section notes for further reading.) One sticking point is the well-known *Arrow Impossibility Theorem*, for which Kenneth J. Arrow won the Nobel Prize in Economics in 1972. It states, in essence, that it is impossible to run a "fair" election if there are more than two candidates on a ballot—or select a "fair" design or attribute if there are more than two from which to choose. There is debate in the design community as to how relevant this theorem is to what designers do, but we believe that the PCC can be used to indicate the collective preferences of a design team.

Suppose a team of 12 designers is asked to rank order three designs: *A*, *B*, and *C*. In doing so, the twelve designers have, as individuals, produced the following twelve sets of orderings:

$$\begin{array}{ll} 1 \text{ preferred } A > B > C & 4 \text{ preferred } B > C > A \\ 4 \text{ preferred } A > C > B & 3 \text{ preferred } C > B > A \end{array} \qquad (3.1)$$

where $>$ is the ranking symbol used to write "*A* is preferred to *B*" as $A > B$.

The collective will of the design team is worked out through the PCC shown in Table 3.3. A point is awarded to the winner in each pairwise comparison, and then the total points earned by each alternative from all of the designers is summed. The aggregate ranking of preferred designs is

$$C > B > A \qquad (3.2)$$

that is, the group consensus was that *C* was ranked first, *B* second and *A* last. Thus, the 12 designers choose design *C* as their collective first choice, though it was not the unanimous first-choice. In fact, only 3 of 12 designers ranked it first. However, as pointed out by Arrow, there is no such thing as a "fair" election, no matter how many voters (assuming at least two), if there are more than two names on the ballot. However, the PCC as applied here provides as good a tool as there is for these purposes, so long as its results are used with the same caution as those of individual PCC's.

How would we use this in a design setting, if not as a direct decision driver? One approach would be to recognize that each design must have had elements or features that were attractive, else the points awarded each would have been far different. For example, had the points awarded in a PCC vote been $C = 24$, $B = A = 6$, we could conclude that the voters didn't see much overall merit in designs *A* and *B*. On the other hand, had the votes turned out $C = 18$, $B = 16$, and $A = 2$, then it would be reasonable to assume that there were two designs perceived to be nearly equal, in which case combining their best features would be a good design strategy.

Pairwise comparison voting by members of a design team must be used carefully because the results can be misleading.

3.3.4   Using subjective values wisely

The pairwise comparison method should be applied in a *constrained*, *top-down* fashion, so that: (1) objectives are compared only when emanating from a common node at the same level of abstraction or level on the objective tree, and (2) the higher-level objectives are compared and ranked before those at lower, more detailed levels. The second point seems only a matter of common sense to ensure that more "global" objectives (i.e., those abstract objectives that are higher up in the objectives tree) are properly understood and ranked before we fine tune the details. For example, when we look at the objectives for the safe

Table 3.3   A collective pairwise comparison chart (PCC) for 12 designers.

Win / Lose	A	B	C	Sum / Win
A	• • • •	1 + 4 + 0 + 0	1 + 4 + 0 + 0	10
B	0 + 0 + 4 + 3	• • • •	1 + 0 + 4 + 0	12
C	0 + 0 + 4 + 3	0 + 4 + 0 + 3	• • • •	14
Sum / Lose	14	12	10	• • • •

ladder (see Figure 3.1), it is more important to decide how we rank safety against marketability than its use for electrical work or for maintenance. Similarly, for the beverage container (see Figure 3.2), it is again more meaningful to rank safety against sales promotion before worrying about whether the container is easier to open than it is harder to spill.

In addition, given the subjective nature of these rankings, when we use such a ranking tool, we should ask whose values are being assessed. Marketing values could easily be included in different rankings, as in the ladder design, for example, where the design team might need to know whether it's "better" for a ladder to be cheaper or heavier. On the other hand, there could be deeper issues involved that, in some cases, may touch upon the fundamental values of both clients and designers. For example, let us turn to the beverage container design, now with the thought of looking at how the design objectives might be ranked if designs were being developed for the two competing companies, GRAFT and BJIC. We show the pairwise comparison charts for the GRAFT- and BJIC-based design teams in Figures 3.4 (a) and (b), respectively. It is clear from these two charts and the scores in their

Goals	Environ. Benign	Easy to Distribute	Preserve Taste	Appeals to Parents	Market Flexibility	Brand ID	Score
Environ. Benign	••••	0	0	0	0	0	0
Easy to Distribute	1	••••	1	1	1	0	4
Preserve Taste	1	0	••••	0	0	0	1
Appeals to Parents	1	0	1	••••	0	0	2
Market Flexibility	1	0	1	1	••••	0	3
Brand ID	1	1	1	1	1	••••	5

(a) GRAFT's weighted objectives

Goals	Environ. Benign	Easy to Distribute	Preserve Taste	Appeals to Parents	Market Flexibility	Brand ID	Score
Environ. Benign	••••	1	1	1	1	1	5
Easy to Distribute	0	••••	0	0	1	0	1
Preserve Taste	0	1	••••	1	1	1	4
Appeals to Parents	0	1	0	••••	1	1	3
Market Flexibility	0	0	0	0	••••	0	0
Brand ID	0	1	0	0	1	••••	2

(b) BJIC's weighted objectives

FIGURE 3.4 Pairwise comparison charts for the design of the new beverage container. Here, goals for the product are weighted one against another by designers working for (a) GRAFT and (b) BJIC. The relative values attached to each goal varies considerably in each chart, thus reflecting the different values held by each company.

right-hand columns that the folks at GRAFT were far more interested in a container that would generate a strong brand identity and be easy to distribute than in it being environmentally benign or having appeal for parents. At BJIC, on the other hand, the environment and the taste preservation ranked more highly, thus demonstrating that subjective values show up in pairwise comparison charts and, eventually, in the marketplace.

It is also tempting to take our *ranked* or ordered objectives and place them on a *scale* so that we can manipulate the rankings in order to attach relative attaching weights to goals or do some other calculation. It would be nice to be able to answer questions such as, "*How much more* important is portability than cost in our ladder?" Or, in the case of the beverage container, "*How much more* important is environmental friendliness than durability. A little more? A lot more? Ten times more?" We can easily think of cases where one of the objectives is substantially more important than any of the others, such as safety compared to attractiveness or cost in an air traffic control system, and other cases where the objectives are essentially very close to one another. However, and sadly, there is no mathematical foundation for scaling or normalizing the rankings obtained with tools such as the pairwise comparison chart. The numbers obtained with a PCC are *approximate*, *subjective* views or judgments about relative value or importance. They do not represent strong measurement. Therefore, we should not try to make these numbers seem more important by doing further calculations with them or by giving them unwarranted precision.

Lastly, and in the spirit of the foregoing, it is also tempting to want to integrate scores from the pairwise comparison charts into our objectives trees, to construct *weighted* objectives trees that explicitly show relative scores for every goal and subgoal. However, we would be making the error we have just identified, that is, building an appealing numerical edifice on an unsound (mathematical) foundation. Thus, we would recommend no more weighting of objectives than indicated at the beginning of this section. It makes sense to do a simple PCC for a set of goals or subgoals stemming from a common node, but it would be unsound to distribute scaling and normalizing numbers across an objectives tree in the hope of meaningfully comparing subgoals on different branches of a tree.

3.4 MEASURING ACHIEVEMENT: QUANTIFYING WHAT THE CLIENT GETS

Metrics are used to measure how well objectives are met.

Having determined what our client would consider a *good design* (objectives), we must also take up the question of determining how well a particular design *actually achieves* all these things. The answer to this question parallels Section 3.3 because it is about *metrics*; that is, standards that measure the extent to which a design realizes our objectives. Ideally, a metric gives us an exact gauge of the objective we are concerned with. In practice, we often make difficult choices about what constitutes an appropriate metric, how we can actually apply that metric, and how much it costs to measure the achievement of a design objective.

3.4.1 Steps for developing metrics

We follow a three-step procedure in selecting metrics.

1. Identify both the units and the scale of something that it is appropriate to *measure* about our objective.

2. Identify a means of assessing the value of a design in terms of those relevant units.

3. Evaluate whether this particular measurement and its subsequent evaluation is feasible.

In many cases we will be forced to cycle through this process iteratively, until we come up with a suitable method of measuring the design in terms of the objective. The process is described in more detail below.

Our first step in establishing metrics for a design attribute is to determine the appropriate units that might be applied to the objective we are concerned with. For an objective of low weight for our ladders, for example, we might consider units related to weight or mass, that is, kg., lb., or oz. For an objective of low cost, we would probably want our metric to be the cost measured in currency, that is, $U.S. in the United States. The appropriate "units" for some cases may be general categories, or subjective rankings (e.g., "high," "medium," and "low").

Once we have established appropriate units of measure and scale, we must take the second step of determining *how* to accurately assign a figure or value to a particular design. An important aspect of this step is to assure that the plan for measuring the design's performance is compatible with the type of scale and measure selected in the first step. This could include, for example, laboratory tests, field trials, consumer responses to surveys, focus groups, etc. Given an objective of low weight, we could determine the weight by using a conventional balance scale. Cost, on the other hand, could be quite difficult to measure, unless we know factors such as the manufacturing techniques to be employed, the number of units to be manufactured, and the components to be included in the design. Estimating costs can be a complex and demanding field (that we will return to in Chapter 8). For now, however, let us assume that we can estimate the costs to manufacture and distribute our ladders.

Our third step in assigning metrics is the determination of whether or not the information derived from using a metric is worth the cost of actually performing a measurement. In some cases we will find that the usefulness of the metric is slight when compared to our own resources or to those needed to obtain the measurement. In such cases we can either develop a new metric, find another means for measuring the expensive metric, or look for an alternative way of assessing our design. There may be several metrics available with which we are equally comfortable, in which case we may be able to select a less expensive alternative. In other cases, we may decide to use a less accurate method to assess our designs. As a last resort, we may decide to convert our objective into a constraint, which allows us to consider some designs and reject others. (Remember that constraints allow us to exclude options from the design space, while objectives allow us to choose among alternative designs.)

Consider again our objective that our ladder design be low cost. It may be that the information needed to accurately assess the manufacturing costs for ladders is not available to the design team without a significant and expensive study. An alternative option might be to estimate the manufacturing cost by adding up the costs of the individual components when purchased in given lot sizes. This disregards a number of relevant costs (e.g., assembling components, overhead), but it would allow the design team to distinguish between designs with expensive elements and designs with cheaper elements. Alternatively, the designers might depend upon expert input from the client and simply rank the designs into ordinal categories such as "very expensive," "expensive," "moderately expensive," "inexpensive," and "very cheap." Failing all else, the designers might be forced to reconstruct this into a constraint, such as, "contains no parts costing greater than $20."

Good metrics measure the right thing, have clear units, and are cost effective.

If a good metric is not affordable, consider making the corresponding objective into a constraint.

3.4.2 Characteristics of good metrics

Good metrics have a number of attributes. The first of these, of course, is that the metric should *actually measure the objective* that the design is supposed to meet. Often we find that designers try to measure some phenomenon that, while interesting, is not really on point for the desired objective. If, for example, our objective is to appeal to consumers, then measuring the number of colors on the package may be a poor metric. A second characteristic is that the metric should be capable of the *correct level of accuracy* or *tolerance*. If "low weight" is one of our objectives, measuring to the nearest ton or to the nearest milligram is not appropriate, at least not for a ladder.

A third characteristic of concern is that the metric should be *repeatable*. That is, if others were to conduct the same test or measurement, they would obtain the same results, subject to some degree of random error. This characteristic can be met either by using standard methods and instruments, or, if no such methods are available, by carefully documenting the protocols being followed. It also makes it incumbent upon the design team to use sufficiently large statistical samples where possible. A fourth characteristic of good metrics, which is related to repeatability, is that the outcomes should be expressed in *understandable units of measure*.

Finally, any metric should elicit only *unambiguous interpretation*. That is, we would like to have metrics whose results lead all of the members of the design team (and all other stakeholders) to the same conclusion about the measurement. We certainly don't want to be in the position of arguing among ourselves about the meaning of a measurement of a given metric.

It is clear that judgment is called for in selecting a good metric. The scale and units should be appropriate to the design objectives, and means of measuring them must be available and affordable. In general, good metrics result from careful thought, extensive research, and ample experience—which suggests that the selection of metrics can certainly be enhanced by the synergy derived from a cooperative, well-functioning team.

3.5 SOME NUTS AND BOLTS OF DEFINING THE PROBLEM

In this section we focus on some of the practical issues involved in trying to clarify and articulate what the client and users want from a design, including issues about how objectives (and constraints) trees evolve, who is asked, when are they asked, and how is the information and knowledge handled and tracked.

3.5.1 Questioning and brainstorming

There are two kinds of activities that design teams can initiate, more or less contemporaneously, after they have been engaged by a client to undertake a design job. The first is asking questions of the client(s), and of others who might have varying degrees of interest in the design. These other stakeholders should include potential users and experts in the field. The experts can include people versed in any relevant technology in or other technical aspects, and marketing experts who are familiar with the market of users toward which a design is aimed. We have already identified (in Section 3.1) the kinds of questions that can be asked, but it is perhaps useful to keep in mind that what is intended is a collegial approach to soliciting information, not an adversarial approach that might put a respondent on the defensive.

It is also useful to be prepared when asking questions. If we know what we are looking for, we can guide the conversation and get more information. It also helps ensure that the people who are answering the questions feel their time is not being wasted. This is very important if a program of structured interviewing of experts and/or users is envisioned because such interviews, or similar detailed survey forms, will not produce useful or serious responses *unless* the respondents feel that it's worth their time to answer a lot of questions.

The second activity that design teams can initiate in the problem definition phase is brainstorming. As we noted in Chapter 2, brainstorming is a group effort in which new ideas are elicited, retained, and perhaps organized into some problem-relevant structure. When the team is brainstorming to identify desirable attributes or features, it is very important that its focus not shift to functions and means. While there will inevitably be some "nonresponsive" suggestions, the team should try to stick to the topic at hand, namely identifying goals and objectives, and perhaps constraints. One way a team can do this is for its leader to present statements of ideas with phrases such as, "A desirable characteristic of the object would be . . ." This will serve to remind the others of the team's focus. Of course, our preferred outcome is a list of attributes and characteristics that can be pruned as described in Sections 3.1.2 and 3.1.3. This list can then be refined into an indented list of objectives for the design, or, an objectives tree.

3.5.2   When and how do we build an objectives tree?

When do we build an objectives tree? Right away? As soon as the client has offered us the design job? Or, should we do some homework first and perhaps try to learn more about the design task we are undertaking?

Build an objectives tree early, and modify it often while defining the problem.

There's no hard and fast answer to these questions, in part because building an objectives list or tree is not a mathematical problem with an attendant set of initial conditions that must be met first. Also, building a tree is not a one-time, lets-get-it-done kind of activity. It's an iterative process, but one that surely should begin after the design team has at least some degree of understanding of the design domain. Thus, some of the questioning of clients, users, and experts should have begun, and some of the tree building can go on episodically while more information is being gathered.

One interesting feature of building an objectives tree is logistical in nature. How do we organize all that information, particularly if we're sitting around a conference room table and doing some intensive brainstorming? Surely we'd use blackboards or whiteboards, but how do we do all the clustering and the hierarchical organization while team members are throwing out ideas in a rapid, stream of consciousness fashion? One way is to use Post-It™ notes, which come in various sizes these days, and to simply write up individual notes for each entry on the list or in the tree. The notes can then be pasted on a board or display and later moved around as the team begins to organize the list of design attributes.

Two minor but important points. First, it is important that someone take notes during brainstorming sessions, in order to ensure that all suggestions and ideas are captured. It's always easier to prune out and throw away things than to recapture spontaneous ideas and inspirations. Second, after a rough outline of an objectives tree has emerged, it can be formalized and made to look pretty (and presentable) simply by using any standard, commercially available software package for constructing organization charts or similar graphical displays.

3.6.3 Revised project statements

Share revised problem statements with the client—they just may be right!

We have assumed all along that design projects would be initiated with a relatively brief statement drafted by the client to indicate what he seems to want. All of the methods and outputs we have described in this chapter are aimed at understanding and elucidating these wants, as well as accounting for the wants of other potential stakeholders. As we gather information from clients, users, and other stakeholders, our views of the design problem will shift as we expose implicit assumptions and perhaps a bias toward an implied solution. Thus, it is important that we recognize the impact of the new information we've developed and that we formalize it by drafting a revised problem statement that clearly reflects our clarified understanding of the design problem at hand. We saw such a revised problem statement as one of the emergent products of the beverage container design (viz., Section 3.1.4), and a comparison of the initial and revised problem statements for this project speaks very clearly to the notion of exposing more precisely what the client wants. We will see a similar result in the next section.

3.6 DESIGNING CHICKEN COOPS FOR A GUATEMALAN WOMEN'S COOPERATIVE

In their first engineering course at Harvey Mudd College, *E4: Introduction to Engineering Design*, first-year engineering students are assigned the task of developing a conceptual design for a device or system. The projects are typically done for the benefit of a nonprofit or educational institution, and they provide the students with the insight that good engineering design may be (and is) done in nontraditional, noncorporate settings. The course also stresses the formal design methods we are presenting in this book. In order to illustrate student design within the E4 environment, we now describe the design of a chicken coop for use in the remote Guatemalan village of San Martin Chiquito. The sponsor of this project, Xela-Aid, is a humanitarian organization committed to working with and for people in Guatemala. Xela-Aid has a long history of working with the students in Harvey Mudd's E4 course. Among the E4 projects that have been done for Xela-Aid are designs for greenhouses, playgrounds, and improved methods for carrying burdens up a steep and treacherous mountain roadway.

In the design problem at hand, E4 teams were asked to design a chicken coop that would increase egg and chicken production, using materials that were readily available and maintainable by local workers. The end users were to be the women of a weaving cooperative who wanted to increase the protein in their children's diet in ways that are consistent with their traditional diet, while not appreciably distracting from their weaving. The full problem statement is (an abbreviated version was given in Section 2.5):

The women of this village currently raise chickens in small fenced areas, but the few eggs and chickens produced must be sold at market to supplement the income of the family. Currently, chickens are kept in a small, fenced area (12 ft. × 12 ft.) with nothing but food on the ground and water bowls inside. Many of the eggs become cracked, and the coop floor becomes lower as the waste is swept out with a broom. Currently, the waste is placed in a shallow ditch. With newly purchased land, the women hope to be able to increase production, and have agreed that half of all the eggs produced will be used for their families, the other half, sold. The challenge will be to raise the most chickens/eggs possible in a small space of land, and in the most cost-efficient manner. Cold temperatures will require that the chickens are kept warm, but while electricity is available, relying on it could make the venture too costly. The women would like a low-upkeep system, as well, by which feed and water containers could be used that would require less filling, but in which the water will not putrefy, nor feed become moldy. A method for easy cleaning is also sought which will not erode the floor of the coop.

A reading of the above initial problem statement makes it clear that the design teams had many questions to answer before they could begin to specify the chicken coop's ultimate form. Probably the most pressing of these questions is, "What exactly do the client and the end users want (and need), and what are the most important of the possible answers?" To answer the first part of this question, the students had to undertake research into chicken husbandry (i.e., how to raise chickens), existing designs for chicken coops, and the cultural and climatic environment of Guatemala. In addition, teams had to determine what terms such as "cost-efficient" and "low-upkeep" meant to the client and to the end users. This was accomplished by a combination of library research, web searches, interviews with local (to Harvey Mudd) poultry producers and researchers, and repeated interviews with the client's liaison. The end result of this was that objectives trees and refined client statements were developed.

3.6.1 Objectives trees for the Guatemalan chicken coop

The objectives trees developed by two different teams are shown in Figures 3.5 and 3.6 *as they did them*. As such, they have some errors or problems that are worth further exploration. For example, "non-eroding floor" in Figure 3.5 may be a means or a constraint, but it is not an objective. However, even with their shortcomings, there are a number of interesting points to be made about these two objectives trees. First, the two trees are clearly not identical. While this is not surprising, given that the trees reflect the work of two different teams, it highlights the fact that many of the objectives and constraints that occur to designers are subject to analysis, interpretation, and revision. Thus, it is very important that designers carefully review their findings with the client before proceeding too far in the design process.

A second point to note is that one of the teams has chosen to incorporate the results of its research (e.g., "prevent egg cannibalism" in Figure 3.6), while the other has kept its objectives at a more general level, reflecting primarily what the client's liaison had indicated in personal interviews. The designer often informs and educates the client, providing the client with a better understanding of the problem as the process of clarifying objectives

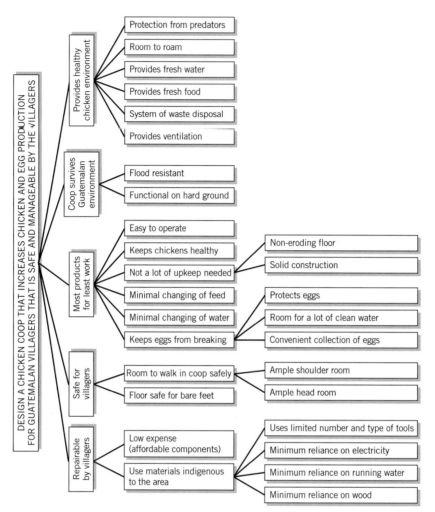

FIGURE 3.5 An objectives tree done by a freshman design team for the Xela-Aid chicken coop project. There are some entries in this tree that don't belong. Can you identify them?

unfolds. This takes on a particular importance when we are considering functions and parametric specifications (discussed in Chapter 6).

3.6.2 Metrics for the Guatemalan chicken coop

The teams developed metrics for their objectives. In the two cases presented here, the teams quickly realized that measuring the performance of designs against some of the objectives was simply beyond the scope of a one-semester, introductory design project. Figure 3.7 shows some of the metrics developed by one team. While these metrics show an interesting grasp of the underlying concerns (e.g., the cigarette test for preventing the spread of fire),

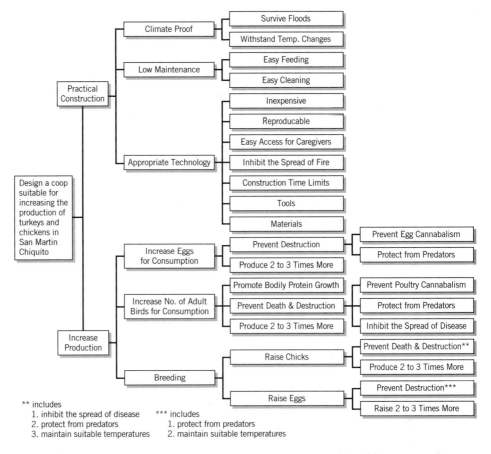

FIGURE 3.6 Another student team's objectives tree for the Xela-Aid chicken coop project. This tree also has some "objectives" that aren't, including a performance specification (see Chapter 4). Note, too, that this team did not consider safety to be a top-level objective.

most of them are beyond the time and other limits to measure and test. As a result, this team essentially abandoned formal metrics for most of the objectives and instead chose a final design by estimating the number of chickens that each coop design could sustain. This is not a recommended approach.

Figure 3.8 shows the simpler metrics and associated tests that another team developed and used. This team elected to test individual components used within their overall design rather than focus on the design itself. Notice that the tests are reduced to simple Yes or No outcomes from simple field tests. This has the virtue of being quite manageable. Unfortunately, it also has the effect of separating the tests from their underlying objectives, thereby making it unclear exactly what is the point of each test. This approach also leads to the rapid adoption of a very small set of design alternatives, within which the selection of components remained somewhat larger. It can be argued that in both cases, the teams might have been more successful had they gone through several thoughtful iterations.

survive floods	simulate flood conditions by using a fire hose on a prototype to see if the footings can be washed away or if the roof leak
withstand temperature changes	attempt to destroy the materials used in the design through repeated freezing and heating (within the constraints of Guatemalan climate)
easy feeding and easy cleaning	ask a volunteer to do the tasks involved and measure the time and effort expended
inexpensive	determine the ratio of net cost to the number of chickens maintained by the coop
reproducible	build a scale model, then give it (and it alone) to a volunteer and ask them to build another one like it out of the same materials
easy access for caregiversor	test for presence, absence of tight (less than three feet wide) corridors high (over five feet) accesses
prevent the spread of fire	test a model's resistance to various types of fires, such as airborne sparks (on a roof), cigarettes dropped in nesting material, etc.
appropriate tools and materials	whether or not they are locally available
protect from predators	build a prototype, put bait inside cages and nest boxes, and then see if any predators can get to it
prevent egg cannibalism	presence of dark nests (based on authoritative references)
increase production	the number of cages and nest boxes available in the design
promote bodily protein growth	credit if chickens have access to the ground to scratch for insects
maintain suitable temperature (for chicks and eggs)	full credit if broody hens are used; if artificial heat is used, test variability of heat source by placing a thermometer near it and measuring the variance of temperature
inhibit spread of disease	make a full prototype with live chickens; infect one chicken with a disease, and track the spread through the population. If a significant number of chickens survive, the design passes.

FIGURE 3.7   This table shows some of the tests (metrics) that were proposed by a student design team for the chicken coop. Many of the tests appear to require either more time or more resources than the team actually had. In this case, the team would have been better served by simplifying the metrics or the tests.

3.6.3   Revised project statement for the Guatemalan chicken coop

After conducting their research and extensive interviews with their client, including several reviews of their objectives trees, the design teams revised their original problem statements

Coop component tested	Materials used in testing	Metrics	Did the component pass the test?
Rebar perimeter fence posts, individual	Rebar posts	Did it support weight when upright?	No.
Bound rebar fence posts	Rebar posts	Did it support weight when upright?	Yes.
Removable tray supports	Rebar	Did it support weight when horizontal?	Yes.
Gutter	Corrugated metal and gutter components	Would the corrugated roof support a gutter?	Yes, using only certain gutter hangers.
Ease of construction and "constructability"	Styrofoam, balsa wood, glue	Were the building instructions clear and complete enough to construct a model?	Eventually.

FIGURE 3.8   This table presents an alternative approach to Figure 3.7. Because of limitations of both time and resources, the team has chosen to evaluate various components in their design, and has converted them to pass/fail tests. Are there similar tests for the objectives shown in Figure 3.6?

and adopted new versions. One of the teams produced the following revised problem statement:

The XELA-Aid organization would like our team to design and produce a chicken coop for use by several families in a small village in Guatemala. The climate of the village is tropical with temperatures rarely dropping below 60°F. The village sits in a valley of mostly porous volcanic soil, is located at a height of approximately 7000 ft above sea level, and is surrounded by rain forest. During the rainy season, which lasts from March through September, the village receives an average of over an inch of rain each day. The chicken coop is to be located on a plot of land 8 ft. wide by 20 ft. long. This segment of land is a portion of a larger plot upon which will also be located a greenhouse, and a building where the women will produce assorted fabrics for sale. Materials that are readily available and which should be considered for use when designing the coop are large segments of chicken wire, rebar, concrete, and locally made concrete blocks. Green wood is also available, but should be used sparingly as it is cut directly from the surrounding forest. Tools and any additional specific materials such as screws, nails, or any other inexpensive compact items will be bought by XELA-Aid but should also be considered scarce when designing the coop. Electricity is available in the village but it is expensive and unreliable and should not be incorporated in the design. The village does not have running water; the water must be carried by foot to the village from a source 30 minutes away. The coop will be constructed by the Guatemalans with the assistance of XELA-Aid members so construction skills will not be an issue as long as the design is kept reasonably simple. The overall goal of the new coop is to allow the families to at least double their current chicken and egg production. This will allow the families to sell half of the chickens and eggs and consume the other half. Secondary goals include minimizing the labor required to maintain the coop and chickens, and utilizing the chicken droppings as fertilizer. To make the coop less labor intensive steps should be taken to make it easy to clean and easy to supply with fresh food and water to minimize spoilage and putrefaction, respectively. The primary predator of the chickens in this part of the country is a small catlike animal able to dig under the current fence protecting the chickens; special precautions should be made to insure this animal is unable to enter the enclosure.

3.7   NOTES

Section 3.1: More examples of objectives trees can be found in (Cross, 1994), (Dieter, 1991), and (Suh, 1990). (Cross, 1994) and (Dieter, 1991) also show weighted objectives trees. The very important notion of *satisficing* is due to (Simon , 1981).

Section 3.2: Constraints are discussed in (Pahl and Beitz, 1997).

Section 3.3: Measurements and scales have recently become controversial in the design community, to a dgree beyond our current scope. Some of the critiques derive from an attempt to make design choices and methods emulate long-established approaches of economics and social choice theory (Arrow, 1951, Hazelrigg, 1996, Hazelrigg, 2001, Saari, 1995, Saari, 2001a and 2001b). Other analyses have taken a more positivist approach (Jones, 1992, and Otto, 1995). The PCC's outlined in the text are exactly equivalent to the best tool offered by the social choice theorists (Dym, Scott and Wood, 2002).

Section 3.5: Insights into the need to find and remove bias and implied solutions in assessing problem statements have been provided by (Collier, 1997).

Section 3.6: The results from the Xela-Aid chicken coop design project are taken from final reports (Gutierrez et al., 1997 and Connor et al., 1997) submitted during the Spring, 1997 offering of Harvey Mudd College's freshman design course, *E4: Introduction to Engineering Design*. The course is described in greater detail in (Dym, 1994b).

3.8   EXERCISES

3.1   Explain the differences between biases, implied solutions, constraints, and objectives.

3.2   The HMCI design team established in Exercise 2.5 has been given the problem statement shown below. Identify any biases and implied solutions that appear in this statement.

> Design a portable electric guitar, convenient for air travelers, that sounds, looks, and feels as much as possible like a conventional electric guitar.

Revise the problem statement so as to eliminate these biases and implied solutions.

3.3   Develop an objectives tree for the portable electric guitar. (Someone will have to play the roles of client and users for this design project.)

3.4   Design a strategy for obtaining the weights for the objectives tree of Exercise 3.3.

3.5   The HMCI design team established in Exercise 2.5 has been given the problem statement shown below. Identify any biases and implied solutions that appear in this statement.

> Design a greenhouse for a women's cooperative in a village located in a Guatemalan rainforest. It would enable cultivation of medicinal preventive herbs and aid the villagers' diets. It would also be used to grow flowers that can be sold to supplement villagers' income. The greenhouse must withstand very heavy daily rains and protect the plants inside. The greenhouse must be made of indigenous materials because the villagers are poor.

Revise the problem statement so as to eliminate these biases and implied solutions.

3.6 Develop an objectives tree for the rainforest project. (Someone will have to play the roles of client and users for this design project.)

3.7 Correct and revise the objectives tree developed by one of the chicken coop teams and shown in Figure 3.5.

3.8 Correct and revise the objectives tree developed by the other chicken coop teams and shown in Figure 3.6.

3.9 Develop a set of metrics for the portable guitar of Exercise 3.2. If a metric is likely to be difficult to measure, indicate how it can be reframed as a constraint.

3.10 Develop a set of metrics for the rainforest project of Exercise 3.5. If a metric is likely to be difficult to measure, indicate how it can be reframed as a constraint.

CHAPTER 4

FUNCTIONS AND SPECIFICATIONS

How can I express what the client wants in engineering terms?

UP TO now we have focused on understanding what the client and users want and need from a design. In this chapter we move from using the language of the client to the language of the engineer. In particular, we want to translate the client and user needs and desires into terminology that helps us realize those needs and measure how well we meet them. In the terminology of engineers and designers, we identify functions and performance specifications. These three terms represent related, yet distinct aspects of how a designed artifact does what it was designed to do.

First we will look at *functions* and functional specifications because they tell us *what the designed object must do* to realize the stated objectives. We have to establish *which* functions have to be performed before we can specify how well the functions must be performed. Then we will consider *performance specifications* that tell us *how well* the designed object must do something.

The ordering of functions, specifications, and metrics (introduced in Section 3.4) is somewhat arbitrary. For example, in many cases a designer will consider how the realization of a particular objective might be measured before she determines precisely the function to be performed. In other cases, some of the performance specifications may have been set out by the client when the design job begins. We present these design tools and techniques in this particular order because it delineates a logical way of thinking about them. However, the iterative nature of design inevitably produces variation in the timing of when functions, performance specifications, and metrics are introduced and processed.

4.1 IDENTIFYING FUNCTIONS TO REALIZE OBJECTIVES

If a child is asked what a bookcase does, he might answer that "It doesn't *do* anything, it just sits there." An engineer, however, would say that the bookcase does a number of different things, and that it must do them well to be a successful design. In this view the bookcase resists the force of gravity exactly, so that books neither fall to the floor nor are forced into the air. The shelves of the bookcase may have dividers, to aid in separating the books into categories chosen by the owner. And, if the bookcase was designed with an eye to interior design, we might say that it enhances the visual appeal of the room. Each of these ways that our designed bookcase does things are functions. An engineer looking at designed objects is educated to see that ~~artifacts do things~~, even when they "just sit there." Understanding what a designed object must do is essential to creating a successful design. In this section we will look more deeply at what we mean when we talk about a design doing something, and then we will look at techniques for understanding and listing functions.

It is particularly important for a designer to be able to properly specify functions because there are consequences for the engineer when he fails to understand and design for *all* of the functions in a design. The literature of forensic engineering is rich with cases in which engineers failed to realize some additional function(s) that had not been met, often with tragic results.

4.1.1 What are functions?

We can think of functions in a number of different ways. In elementary calculus, for example, we say that y, a dependent variable, is a *function* of x, a single independent variable, when we write $y = f(x)$. That is, the value of y depends on the value of x. In multivariate calculus we extend this notion to include multiple independent variables and multiple dependent responses. Management studies refer to *transformation functions* in which a vector of inputs (labor, materials, technology, etc.) is transformed into a set of outputs (products, services, etc.). In each of these cases we are highlighting the existence of a relationship between some independent variables (i.e., *inputs*), and response or dependent variables (i.e., *outputs*), and we are characterizing that relationship in a formal way. We use a similar notion of functions when we consider the functions of an object we are designing. To designers, functions are the things that the designed object must do in order to be successful. As such, the statement of a function usually consists of an "action" verb and a noun. For example, lift, raise, move, transfer, or illuminate are action verbs. The noun in the statement of function may start off as a very specific reference, but experienced designers look for more general cases.

Functions are often expressed as verb-object pairs.

For example, one of the bookcase functions was to resist forces of gravity, which we might characterize as "support books." However, this would imply that the bookcase would only be used to hold books. But shelves in bookcases often support trophies, art, or even piles of homework. Thus, a more basic and more useful statement of the function to be served here is that shelves resist the force of gravity associated with objects weighing less than some predetermined weight. That is, our statement of function is that shelves will support some number of kg. (or lbs.). When describing functions, then, we should use a verb-noun combination that best describes the most general case.

We also want to avoid tying a function to a particular solution. If we were designing a cigarette lighter, for example, we might be tempted to consider "applying flame to tobacco" as a function. This implies that the only way to light the tobacco is by using a flame (and that tobacco is the only material to be lit). Car lighters, however, use electrical resistance in a wire to achieve this function. Thus, a better statement of this function might be "ignite leafy matter," or even "ignite flammable materials." (Somewhat parenthetically, we could consider the following questions. In light of the well-documented health hazards associated with smoking, is there an ethical issue for an engineer who is asked to design a better cigarette lighter? Is it an appropriate design task? We discuss ethics in engineering and design in Chapter 9, but we note here that the issue may not arise if the lighter is viewed as a camping tool.)

We can also categorize functions as being either *basic* or *secondary* functions. A *basic function* is defined as "the specific work that a project, process, or procedure is designed to accomplish." *Secondary functions* would be (1) any other functions needed to do the basic function or (2) those that result from doing the basic function. Secondary functions can themselves be either required or unwanted functions. *Required secondary functions* are clearly those needed for the basic function. Consider, for example, an overhead projector. Its basic function is to project images, and it has required secondary functions that include converting energy, generating light, and focusing images. *Unwanted secondary functions* are undesirable byproducts of other (basic or secondary) functions. For the overhead projector these include generating heat and generating noise. Such undesirable byproducts often generate new required functions, such as quieting noise or dissipating generated heat.

4.1.2   How do we identify and specify functions?

How do we identify and specify the functions to be performed by the artifact that is the focus of our particular design project? We must determine these functions in order to ensure that our final design does what it is supposed to do. Thus, we now describe four methods used to determine functions: enumeration, analysis of "black" and "transparent" boxes, construction of function-means trees, and reverse engineering, a method that is also known as dissection.

4.1.2.1 Enumeration   The most basic method of determining functions for a designed object is to simply *enumerate* or list all of the functions that we can readily identify. This is an excellent way to begin functional analysis for many objects. It leads us to consider what the basic function of the object is and it may prove useful for determining secondary functions. However, we might get "stumped" very early in this process. Consider, for example, a bridge. If the bridge is used for highway traffic, we might note that its basic function is to act as a conduit for cars and trucks, and then we might scratch our heads before being able to add much to this initial, single-entry list. However, there some useful "tricks" we can use to extend an enumerated list.

Effective function enumeration goes beyond making a list.

One trick is to imagine that an object exists and ask what would happen if it suddenly vanished. If a bridge disappeared entirely, for example, any cars on the bridge would fall into the river or ravine over which the bridge crosses. This suggests that one function of a bridge is to support loads placed on the bridge. If the abutments ceased to exist, the deck and superstructure of the bridge would also fall, which suggests that another function of the bridge is to support its own weight. (This may seem silly until we recall that there have been more than a few disasters in which bridges collapsed because they failed to support even their own weight during their construction. Among the most famous of such infelicitous bridges is the Quebec Bridge over the St. Lawrence River, which collapsed once in 1907 with the loss of 75 lives and again in 1916 when its closing span fell down.) If the ends of a bridge that connect to various roads disappeared, traffic would not be able to get on the bridge, and any vehicles on the bridge would be unable to get off. This suggests that another function of a bridge is to connect a crossing to the road network. If road dividers on our bridge were removed, vehicles headed in one direction can collide with vehicles headed in the other. Thus, separating traffic by direction is a function that many bridges serve, and it is a function that can be accomplished in several ways. For example, New York's George Washington Bridge assigns different directions of traffic on each its two levels. Other bridges use median strips.

Another way to determine functions is to consider how an object might be used and maintained over its lifetime. In the case of our bridge, for example, we might note that it is likely to be painted, so that one function is to provide the bridge's maintenance workers with access to all parts of the structure. This function might be served with ladders, catwalks, elevators, and so on.

Consider once again our beverage container design problem. Here, because we have ample experience with such containers, we can readily name or list the functions served by a beverage container, including at least the following:

- contain liquid
- get liquid into the container (fill the container)
- get liquid out of the container (empty the container)
- close the container after opening (if it is to be used more than once)
- resist forces induced by temperature extremes
- resist forces induced by handling in transit
- identify the product

Note that the functions of getting liquid into and out of the container are distinct. This is evident after a brief reflection on canned beverages: Liquid is sealed in by a permanent top, while access is obtained through a pull tab. We might have noticed this distinction between the filling and emptying functions had we considered the "life cycle" of a beverage container.

At the heart of our approaches to function enumeration lies the need for the designer to list the verb-noun pair that corresponds to each and every function of the designed object. However, since enumeration is often difficult, we must turn to other methods.

4.1.2.2 Black boxes and transparent boxes

Black box analyses connect outputs to inputs.

Recall that our earlier discussions of mathematical and management functions featured inputs and outputs, both singly or in groups. This input-output model is also useful in modeling system designs and their associated functions. One tool that helps relate inputs and outputs, and the transformations between them, is the *black box*. A black box is a graphic representation of the system or object being designed, with inputs shown entering the box on its left-hand side and outputs leaving on the right. *All* of the known inputs and outputs should be specified, even undesirable byproducts that result from unwanted secondary functions. In many cases functional analysis helps us identify inputs or outputs that have been overlooked. Once a black box has been drawn, a designer can ask questions such as "What happens to this input?" or "Where does this output come from?" We can answer such questions by removing the cover of the black box, thus making it into a *transparent box*, to see what is going inside. That is, we expose transformations from inputs to outputs by making a box transparent. We can also link more detailed "subinputs" to (smaller) internal boxes within any given box that produce related "suboutputs."

Consider, for example, a system with three inputs: an airborne signal within that part of the frequency spectrum containing the radio frequencies (RF), a controllable source of electrical power, and a vector of desired outputs (such as particular stations and volume levels). This system has three obvious outputs: sound, heat, and a display that indicates how the user's desired frequency and volume level were realized. This system, the radio, can be viewed as being contained in a box that transforms an RF signal into another signal that we hear the audible sounds of music, talk, and perhaps noise! How does this happen? What functions are performed in a radio? Can we identify the many functions of a radio by looking inside the radio box?

We model the functions of our radio with the series of boxes shown in Figure 4.1. In Figure 4.1(a) we see the radio's most basic function of converting an RF signal to an audio signal is achieved while several byproducts, both desired and undesired, are generated at the same time. If we take the cover off this box (Figure 4.1(b)), we see several new black boxes within. These boxes include transforming the power from that of a wall outlet, 110V, to a level appropriate for the radio's internal circuitry, probably 12V. Other internal functions include filtering out unwanted frequencies, amplifying the signal, and converting the RF to an electrical signal that drives speakers. Thus, making our black box cover transparent revealed a number of additional functions. If we had to design the radio, we would probably remove the covers of even more of the boxes we now see. If, on the other hand, we were assembling a radio from known parts, we might stop at this level. This method of making internal boxes transparent and analyzing their internal functions is also called the *glass box method*. No matter which term we use, the effect is the same: We keep opening internal boxes until we fully understand how all inputs are transformed into corresponding outputs and what additional side effects are produced by these transformations.

> Black boxes become glass boxes when we ask *how* inputs produce outputs.

The black box method can be a very effective way to determine functions, even for systems or devices that do not have a *physical* box or housing. The only requirement for using a black or transparent box is that *all* of the inputs and outputs be identified. For example, to design a playground for a rainy climate, our inputs would include children, their parents or caregivers, and the rain. Our outputs would include entertained children, satisfied parents, and water. If we forget the water, our playground design will suffer for lack of proper drainage. (As an aside, it is generally not sufficient to include general terms such as "weather" unless we are willing to consider how the weather is translated into water, ice, wind, and heat *within* our box.)

A final point on the black or glass box method is that we must be very careful when we define the *boundaries* or limits of the device or system for which we are identifying functions. These boundaries require a tradeoff. If they are too wide, we may specify functions that are beyond our control, for example, specifying the household electric current for the radio. If

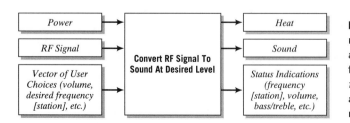

FIGURE 4.1 (a) This is a black box for the radio. Notice that all the inputs and outputs are somehow related in the single, top-level function. We call this top-level function a *basic function*. If we want to know how the inputs are actually transformed into the outputs, we need to remove the cover on the black box.

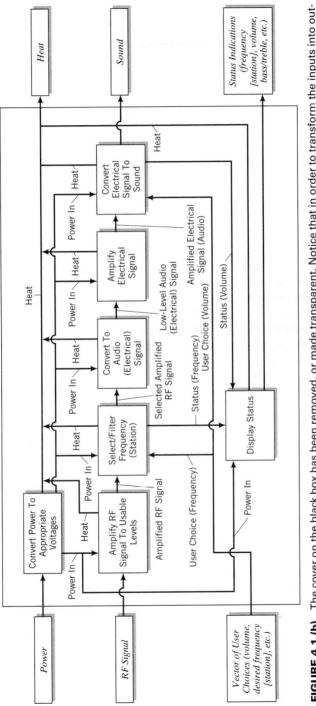

FIGURE 4.1 (b) The cover on the black box has been removed, or made transparent. Notice that in order to transform the inputs into outputs, a large number of secondary functions (or our curiosity) demanded it, we could also remove the covers on some or all of these functions. Notice also that this design calls for the heat to be allowed out of the box on its own. In many designs, we would add a specific function, "dissipate heat," and then decide on a strategy for doing so.

boundaries are drawn narrowly, they may limit the scope of the design. For example, it may be an open issue whether the radio output is an electrical signal that is fed to the speakers or whether the radio output is the acoustical signal coming from the speakers. The question being decided by the placement of the boundary, then, is whether or not the speakers are part of the radio. Such decisions are generally resolved by the client and potential users.

4.1.2.3 Function-means tree

We often have ideas about how a designed device or system might work early in the design process. While we warn against "marrying your first design" and caution against trying to solve design problems until they are fully understood, it is often true that early design ideas suggest different functional aspects. Consider the hand-held (cigarette) lighter. Clearly, if we use a flame to ignite leafy materials we will encounter different secondary functions than if we use hot wires or lasers. One such difference could be the shielding of the igniting element if the device is to be pocket-sized. A function-means tree can help us sort out secondary functions in cases where means or implementations can lead to different functions.

A *function-means tree* is a graphical representation of a design's basic and secondary functions. The tree's top level shows the basic function(s) to be met. Each succeeding level alternates between showing means by which the primary function(s) might be implemented and displaying the secondary functions made necessary by those means. Some graphical notation is employed to distinguish functions from means. For example, functions and means can be shown in boxes with different shapes or written in different fonts. Figure 4.2 shows part of a function-means tree for the hand-held cigarette lighter. Note that the top level function has been specified in the most general terms possible. At the second level, a flame and a hot wire are given as two different means. These two means imply different sets of secondary functions, as well as some common ones. Some of these secondary functions and their possible means are given in lower levels.

Once a function-means tree has been developed, we can list all of the functions that have been listed, noting which are common to all (or many) of the alternatives and which are particular to a specific means. Functions that are common to all of the means are likely to be inherent to the problem. Others are addressed only if the associated design concept is adopted after evaluation.

A function-means tree has another useful property because it begins the process of associating what we must do with how we might do it. We will return to this issue in Chapter 5 when we present a tool to help us generate and analyze alternatives. That tool, the morphological chart, lists the functions of the designed artifact and the possible means for realizing each function in a matrix format. The effort we put into the function-means tree really pays off then.

Two cautions should be noted regarding function-means trees. The first, and perhaps most obvious, is that a function-means tree *is not* a substitute for either formulating the problem or for generating alternatives. It may be tempting to use the outcome of the function-means tree as a complete description of the available alternatives, but doing so will likely restrict the design space much more than needs be. The second caution is that function-means trees should not be used without using some of the other tools described above. One mistake commonly made by novices (or students) is that they adopt a tool because it somehow "fits" with their preconceived ideas of a solution. This transforms the design process from a creative, goal-oriented activity into just a mechanism for making

Beware of using function-means trees to reinforce preconceived ideas.

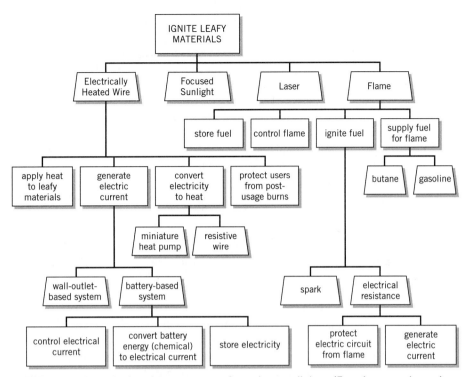

FIGURE 4.2 Part of a function-means tree for a cigarette lighter. (Functions are shown in rectangles, while means are shown in trapezoids.) Notice that there are different subfunctions that result from different means. It is often the case that conceptual design choices will result in very different functions at the preliminary and detailed design stages.

choices that a designer wanted. That is, because the function-means tree allows us to work with appealing means or implementations, we may overlook functions that might have turned up with a less "solution-oriented" technique.

4.1.2.4 Dissection and reverse engineering
Most engineers, and indeed, most curious people, ask the question "What does this do?" when confronted with a button, knob or dial. A natural follow-up may take the forms "How does it do that?" or "Why would you want to do that?" When we follow up these questions with remarks on how we might do it better or differently, we are engaging in the art of *dissection* or *reverse engineering*. Reverse engineering means taking an artifact or device that does some or all of what we want our design to do and dissecting—or deconstructing or disassembling—it to find out, in great detail, just how it functions or works. We may not be able to use that design for any number of reasons: It may not do all the things we want, or do them very well; it may be too expensive; it may be protected by a patent; or it may be our competitor's design. But even if all of these reasons apply, we often can gain insight into our own design problem by looking at how other people have thought about the same or similar problems.

The process is actually quite simple. We begin with a means that has been used by a previous designer, then determine what functions are realized by that given means. We then explore alternative ways of doing the same thing. For example, to understand the functioning of an overhead transparency projector, we might find a button that, when pushed, turns on the projector. A projector button controls the function of turning the projector on or off. This can be done in other ways, including toggle switches and bars along the front of the projector. It is an interesting exercise to consider just how many functions can be thought of for this commonplace device.

Several cautions should be noted about using reverse engineering to find functions. First, the devices being dissected were developed to meet the goals of a particular client and a target set of users. This audience may have had very different concerns than are called for in the current project. Thus, a designer should be sure to stay focused on the current client's needs. Second, there is often a temptation to limit the new means to those that work in the context of the object that is being dissected. For example, all of the means for turning on and off the power to the classroom projector are more or less compatible with a stand-alone device. In some settings, however, it might be more appropriate to remove these controls from the device itself and make them part of some more general room controls. In theaters, for example, lights and other controls are often located in the projection room, rather than in wall switches. It is important that we do not become captive to the design being used to assist our thinking.

A third caution is that while we treat the terms dissection and reverse engineering as equals, they may not always refer to exactly the same process. This is because dissection is sometimes viewed just as it is in a high school biology laboratory, wherein a frog is dissected to reveal its anatomical structure. Here, dissection is more descriptive than analytical. In reverse engineering we go a step further as we try to determine means for making functions happen, which means that we are trying to analyze both the functional behavior of a device and how that functional behavior is implemented.

There is a fourth consideration, one that we stated earlier but reiterate here. We need to define functions in the broadest possible terms and only focus down when it is necessary. Restricting functions to the most immediate terms found on the object being reverse engineered may lead us to mimic someone else's design, rather than fully appreciating opportunities for new ideas. Further, there are clearly serious intellectual property and ethical issues tied to reverse engineering. It is never appropriate to claim as our own the ideas of others. In some cases this can be a violation of law. We will discuss intellectual property in Chapter 5 and ethics in Chapter 9, but it is always important to respect the ideas of others at least as stringently as any other (tangible) property they might hold. After all, wouldn't we want the same protections for our own ideas?

> Dissection should enhance our appreciation of the ideas of others.

4.1.3 A repeated caution about functions and objectives

It is often the case that novice designers make lists of objectives when functions are called for, and vice versa. This happens because two new concepts—objectives and functions—are being learned, and the tools used to determine objectives and functions, respectively, are sufficiently similar that they are often confused. The "newness" problem will likely be overcome only by listing examples of each concept and discussing them with team members, teachers, and experienced designers, hoping to obtain in this way a practical feel for both objectives and functions. The confusion related to the tools themselves can be eased by keeping in mind whether the immediate focus is on "being" terms or on "doing" verbs. As we noted in Section 3.1.2:

Objectives are adjectives; functions are verbs.

- Objectives describe what the designed artifact will be like, that is, what the final object will be and what qualities it will have. As such, objectives detail attributes and are usually be characterized by present participles such as "are" and "be."
- *Functions* describe what the object will *do*, with a particular focus on the input-output transformations that the artifact or system will accomplish. As such, *functions transform inputs to outputs*, and are usually characterized by active verbs.

The distinction between objectives and functions is terribly important, but its centrality is often only grasped fully after a great deal of serious practice.

4.2 DESIGN SPECIFICATIONS: EXPRESSING ATTRIBUTES AND FUNCTIONAL BEHAVIOR

In Chapter 1 we mentioned that design specifications articulate in various ways the attributes and behaviors of a design. Such specifications also provide a basis for evaluating a design because such "specs" become the targets of the design process against which we measure our success in achieving them. Design specifications or requirements are presented in three forms that represents three ways of formalizing what the client or user wants in terms suitable for engineering analysis and design:

*Prescriptive specifications specify **values** for attributes of the designed object.* For example, "A step on a ladder is safe if it is made from Grade A fir, has a length that does not exceed 20 in., and is attached in a full-width groove slot at each end."

*Procedural specifications identify specific **procedures** for calculating attributes or behavior.* For example, "A step on the ladder is safe if its maximum bending stress is computed from $\sigma_{max} = Mc/I$ and is such that σ_{max} does not exceed σ_{allow}."

*Performance specifications identify **performance levels** that signify the achieved desired functional behavior.* For example, "A step on a ladder is safe if it supports an 800 lb. gorilla."

Thus, *prescriptive* specifications specify values of attributes that a successful design must meet (e.g., a chicken coop should have a clear plastic roof). *Procedural* specifications mandate specific procedures or methods to be used to calculate attributes or behavior (e.g., the ability to house chickens safely might be specified by requiring application of the USDA poultry guidelines.) .

Performance specifications characterize the desired functional behavior of the designed object or system (e.g., a chicken coop should house 25 chickens). As noted in Section 4.1, determining what a designed object or system must do is essential to the design process. Functional specifications don't mean much if we don't consider *how well* the design must perform its functions. For example, if we want a device that produces musical sounds, we should specify how loudly, how clearly, and at what frequencies the sounds are produced. Thus performance or functional requirements must be specified or defined.

In addition, if a system or artifact has to work with other systems or artifacts, then we must specify how those systems interact. We call these particular specifications *interface performance specifications*.

4.2.1   Attaching numbers to design specifications

It is normally up to the designer to cast functions in terms that facilitate the application of engineering principles to the design problem at hand. Thus, the designer has to translate functions into measurable terms in order to be able to develop and assess a design. We must find a way to measure the performance of a design in realizing a specific function or objective and then establish the range over which that measure is relevant to the design. And designers must determine the extent to which ranges of improvements in performance really matter.

Determining the range over which a measure is relevant to a design and deciding how much improvement is worthwhile are interesting problems. Our conceptual starting point for assessing the value of a gain in performance of a design (at some unspecified cost) is the curve shown in Figure 4.3. It is similar to what economists call a *utility plot* with which the benefit of an *incremental* or *marginal* gain in performance can be found. The utility or value of that design gain is plotted on the ordinate on a normalized range from 0 to 1. The level or "cost" of the attribute being assessed is shown on the abscissa. For example, consider using processor speed as a measure of a laptop computer performance. At processor speeds below 100 MHz, the computer is so slow that a marginal gain from, say, 50 MHz to 75 MHz, provides no real gain. Thus, for processor speeds below 100 MHz, the utility is zero. At the other end of the utility curve, say, above 5 GHz, the tasks for which this computer is designed cannot exploit additional gains in processor speed. For example, browsing the World Wide Web may be more constrained by typing speed or communication line speeds, so that an incremental gain from 5 GHz to 5.1 GHz still leaves us with a normalized utility of 1. Thus, the utility plot is *saturated* at high speeds.

What happens at performance levels between those that have no value and those on the saturation plateau, say between 100 MHz and 3 GHz for the computer design? In this range we expect that changes do matter, and that increases in processor speed do improve the incremental or marginal gain. In Figure 4.3 we show an *S*- or Saturation-curve that dis-

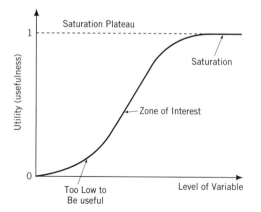

FIGURE 4.3   A hypothetical performance specification curve. Notice that until some minimal level is realized, no meaningful benefit is achieved. Similarly, above some saturation plateau, there is no meaningful benefit in gaining still more. The actual shape of the curve is likely to be uncertain in most cases.

plays *qualitatively* what is happening. There are clearly gains to be made as we move toward faster speeds, and the value of those gains can be determined from the curve. Thus, the utility of the entire S-curve is initially flat (or 0) at low processor speeds, increases measurably over a range of interest, and then plateaus at 1 because added gains are not valued.

This sort of behavior seen in a utility plot is rather common. Economists refer to the *law of diminishing returns*, however, it isn't a "law." We don't usually know the actual shape or precise details of the S-curve (it may not look nearly as smooth as what we have sketched in Figure 4.3), so we choose to approximate it by a collection of straight lines, such as that shown in Figure 4.4. Here there are still regions where gains no longer interest us, as indicated by the horizontal lines at levels 0 and 1. In the middle range, however, we suppose that we are just as happy to increase our levels of the design variable (e.g., processor speed) to obtain a corresponding linear gain in the utility, anywhere within this range of interest. Perhaps most important here is the point that, qualitatively, we are simply saying that the straight line defines a range within which we expect to achieve design gains by tuning the design variable in question.

Consider another example. Suppose we are asked to design a Braille printer that is quiet enough to be used in office settings. None of the competing designs are quiet enough to be so used. How quiet does this design really have to be? To answer this question, we must determine the relevant units of noise measurement and the range of values of these units that are of interest. We would also find out how much noise is generated by current printer designs and whether or not listeners can distinguish among different designs. If one printer produces the same noise level made by a pin dropping on a carpet, while another generates the noise level of a ticking watch, we would likely view both as quiet enough to be fully acceptable. Similarly, if one printer is as loud as a gas lawn mower, and another is as loud as an unmuffled truck, there is no utility gained by distinguishing between these two designs as neither design will be used in an office setting. (Note that this example shows a *reverse* S-curve in which we start at saturation because there is no gain to be made at such low levels of quietness, and then we degrade to a level of no utility for printers that are uniformly too loud.)

Since sound intensity levels are usually measured in decibels (dB), we might conclude that some range of dB is likely to be of interest. Carrying this further, we might look

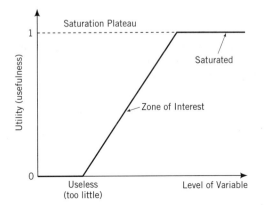

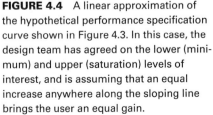

FIGURE 4.4 A linear approximation of the hypothetical performance specification curve shown in Figure 4.3. In this case, the design team has agreed on the lower (minimum) and upper (saturation) levels of interest, and is assuming that an equal increase anywhere along the sloping line brings the user an equal gain.

TABLE 4.1 **Sound intensity levels that are produced by various devices and are measured in various environments. Sound intensity levels are measured in decibels (dB) and are a logarithmic expression of the square of acoustic power. Thus, a 3 dB shift corresponds to a doubling of the energy produced by the source, while the human ear cannot distinguish between levels that differ by only 1 dB (or less). (Adapted from Glover, 1993.)**

Level (dB)	Qualitative Description	Source/Environment
10	Very Faint	Hearing Threshold; Anechoic Chamber
20	Very Faint	Whisper; Empty Theater
30	Faint	Quiet Conversation
40	Faint	Normal Private Office
50	Moderate	Normal Office Background Noise
60	Moderate	Normal Private Conversation
70	Loud	Radio; Normal Street Noise
80	Loud	Electric Razor; Noisy Office
90	Very Loud	Band; Unmuffled Truck
100	Very Loud	Lawn Mower (Gas); Boiler Factory

for some indication of how much noise is produced by other devices and within different environments. We show sound intensities for various devices and environments in Table 4.1. For reference, we show the noise exposure levels to which workers may be exposed in Table 4.2. These levels, expressed in hours of exposure, are defined by OSHA, the federal agency concerned with the safety of work environments. With such environmental and exposure information in hand, the designer can identify a range of interest for a performance specification for the Braille printer. New printer designs must generate less than 60 dB noise in an office environment. Further, smaller values of generated noise levels are considered gains, down to a level of 20 dB. All designs that generate less than 20 dB are equally good. All design that produce more than 60 dB are unacceptable. Note that any realistic designs will generate noise levels that are so far below the OSHA exposure values that occupational safety is not an issue here.

TABLE 4.2 **Permissible noise exposures in American work environments, expressed in intensity levels (dB) permitted during various daily durations (hours). These levels and durations are defined by the Occupational Safety and Health Administration (OSHA), If workers are exposed to levels above these or for period of times longer than indicated, they must be given hearing-protection devices. (Adapted from Glover, 1993.)**

Daily Duration (Hours)	Sound Level (dB)
0.5	110
1	105
2	100
3	97
4	95
8	90

4.2.2 Setting performance levels

Performance
specifications
require sound
engineering,
reasonable
measurements,
and clarified
client interests.

We now extend the above discussion to setting performance levels. First, we determine design parameters that reflect the functions or attributes that must be measured and the units in which those parameters are to be measured. Then we establish the range of interest for each design parameter. For desirable design variables (i.e., qualities or attributes), utility values *below a threshold* are treated as equals because no meaningful gains can be made. Utility values above a *saturation plateau* are also indistinguishable because no useful gains can be achieved. (We are assuming a standard S-curve in which the threshold comes first and the plateau last.) The range of interest lies between the threshold and the plateau. It is within this zone that the design gains should be matched to and measured with respect to the design parameters that are the subject of a given performance specification. This process works well when we exercise judgment in setting performance specifications based on: sound engineering principles, an understanding of what can and cannot be reasonably measured, and an accurate reflection of both client's and users' interests.

Consider once again the beverage container. Each of the functions that were specified in Section 4.1.2.1 has a range of values that must be specified. Some of those functions and some relevant questions associated with each function are:

- *Contain liquid*: How much liquid must the container contain and at what temperatures? Is there a range of fluid amounts that we can put into a container and still meet our objectives?

- *Resist forces induced by temperature extremes*: What temperature ranges are relevant? How might we measure the forces created by thermal stresses on the container designs?

- *Resist forces induced by handling in transit*: What are the range of forces that a container might be subject to during routine handling? To what degree should these forces be resisted in order for the container to be acceptable?

Note that similar but distinct problems arise for the second and third functions on this list, as they both relate to forces.

We can now develop a set of performance specifications that the container designs should meet by addressing these and similar questions. For example, we might indicate that each container must hold 12 ± 0.01 oz. In this case the specification has become a constraint because the corresponding utility plot is a simple binary switch: We either meet this design specification or we don't. (Of course, it is possible to study the container design problem as one in which a variable single-size serving is possible, in which case there may exist a linearized S-curve for the container size where smaller is better.) Still another performance requirement could emerge from a production concern, namely, that the containers can be filled by machines at a rate of 60–120 containers per minute. Thus, any container that cannot be filled at least this fast creates a production problem, while a faster rate might exceed current demand projections.

We might also specify that the designs should allow the filled containers to remain undamaged over temperatures from –20 to +140°F. Temperatures lower than a threshold of –20°F are unlikely to be encountered in normal shipping, while temperatures higher than a plateau of +140°F indicate a storage problem. It may be that some designs that appeal in other ways are limited here by either temperature extreme. A judgment will have to be made about the importance of this function and its associated performance requirement.

It is also worth noting that the specification of the performance of a device is often published *after* it has been designed and manufactured because users and consumers want to know whether the product is appropriate for *their* intended use. End users, however, are usually not parties to the design process, and so they depend on published performance specifications that set out the performance levels that can be expected from a device or system. In fact, in many instances designers examine the performance requirements of similar or competing designs to gain insight into issues that may affect end users.

4.2.3 Interface performance specifications

As previously noted, performance specifications also specify how devices or systems must work together with other systems. Such requirements, called *interface performance specifications*, are particularly important in cases where several teams of designers are working, on different parts of a final product, and all of the parts are required to work together smoothly. For example, a designer must ensure that the final design of a car radio is compatible with the space, available power, and the wiring harness of the car. Thus, a design team that has divided a project into several parts must ensure that the final parts will work together. In such cases, the boundaries between the subsystems must be clearly defined, and anything that crosses the boundaries must be specified in sufficient detail to allow all teams to proceed.

Interface performance specifications are increasingly important for large firms that, in an increasingly competitive international arena, are trying to minimize the total time needed to design, test, build, and bring to market new products. Most of the world's major automobile companies, for example, have reduced their design and development times for new cars to one-half or less of what they were a decade ago by having design teams work *concurrently*, or at the same time, on many systems or products, all of which must work together and be suitable for manufacture. This puts a premium on the ability to understand and work with interface performance specifications.

Developing interface performance specifications is easy theoretically, but is extremely hard in practice. During conceptual design, the boundaries or interfaces between the systems that must work together must be specified, and then specifications for each item that crosses a boundary must be developed. These specifications might be a range of values (e.g., 5 V, ± 2 V), or logical or physical devices that enable the boundary crossing (e.g., pinouts, physical connectors), or simply an agreement that a boundary cannot be breached (e.g., between heating systems and fuel systems). In every case the designers of systems on both sides of a boundary must have reached a clear agreement about where the boundary is and how it is to be crossed, if at all. This part of the process can be difficult and demanding in practice since teams on all sides are, in effect, placing constraints on all of the others. A black box functional analysis could be helpful in developing interface specifications because it allows all of the parties to identify the inputs and outputs that must be matched and to deal with any side effects or undesired outputs.

4.2.4 On metrics and performance specifications

The distinction between metrics (defined in Section 3.4) and performance specifications is often a source of confusion. Both involve quantifying or otherwise specifying how well a design does something. In some cases, metrics that are adopted for an objective may also be used as the specifications for some functions. Nevertheless, there are important differences to keep in mind:

Metrics measure objectives; performance specs are measures of functions.

- *Metrics apply to objectives (only).* They allow both designers and clients to assess the extent to which an objective has been realized by a particular design.
- *Performance specifications are typically applied to functions.* They specify how well those functions must be realized by a design.

Performance specifications also take on something of the characteristics of constraints since any design that fails to meet a performance requirement can be considered to have failed. Metrics are needed for *all* of the objectives that are being considered in the design selection process; performance specifications apply only to functions for which there are definable limits of acceptability.

4.3 FUNCTIONS FOR THE XELA-AID CHICKEN COOP DESIGN

In Chapter 3 we clarified and documented the objectives for the Xela-Aid chicken coop design project and refined the problem statement. Here we present some functional analyses and performance specifications that the student teams developed. Remember that we are presenting student design work to demonstrate the largely successful application of the principles and techniques we are describing. The ability to review design documents is another important design skill that can be honed by reading and evaluating the design efforts of other terms, albeit with patience and prudence.

A chicken coop performs a number of functions. One design team used the enumeration method to develop the list of functions shown in Table 4.3. Notice that many of the functions are of the form "allow...." While this is a reasonable way to begin functional

TABLE 4.3   One list of functions for the Xela-Aid chicken coop, as developed by one of the freshman design teams.

Protect chickens from predators during the day
Protect chickens from predators during the night
Protect chickens from weather
Allow for feeding chickens
Allow for watering chickens
Keep water fresh
Allow for egg collection
Protect eggs
Allow incubation of eggs
Ventilate coop
Allow for nest cleaning
Allow for chicken entry
Allow for waste removal
Allow human entry/exit from coop
Allow human entry/exit from perimeter

analysis, a function can often be cast in a more active form. For example, "allow for removal of waste" might be stated as "remove waste." Similarly, "allow for chicken feeding" can be more succinctly stated as "feed chickens." Other functions listed in Table 4.3 suggest solutions. For example, "keep water fresh" implies that water should be changed frequently. There are, of course, alternative solutions, including removing old water, streaming water continuously, and detoxifying the water. A more general statement of function might be "supply chickens with potable water." We are not nitpicking; it is important that designers specify functions in general terms that do not imply solutions. If we don't do that even for relatively simple problems, we can expect to encounter difficulties in more complex design problems whose solutions are far less obvious.

In Figure 4.5 we show the black box analysis of a chicken coop done by another student team. The tables below the diagram list some of the inputs and outputs and their roles in various aspects of coop use. Such tables help designers to identify what must go into and what must come out of a black box. The black box in Figure 4.5 has a number of interesting features and a few problems. Notice that the team has determined that "spoilt water"—which is an object, not a true function—is an output of the water container. This allows the team to identify both a function, "separate spoilt water," and an output, "bad water." There are several functions that have no obvious outputs, such as sleep and exercise, suggesting that the design team has confused the functions of the coop with the functions of the coop's residents. Further review of this figure will no doubt reveal other problems and errors with this example.

The teams responsible for developing the chicken coop designs were not required to develop performance specifications for their designs. While this leaves a hole in the documentation, it also leaves an interesting mental challenge or thought exercise. For example, what are some of the prescriptive design specifications that could apply? How might they be derived? What is the interface between a chicken coop and the surrounding area?

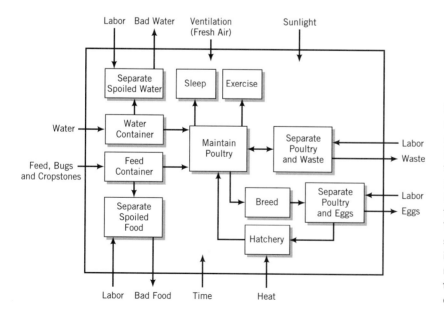

FIGURE 4.5 A black-box analysis of the Xela-Aid chicken coop as developed by one of the student teams. There are several problems with this black box. One is that some of the boxes have no outputs. This seems unlikely, and should raise questions immediately. More fundamentally, it is not clear if this is a black box for the chickens or for the coop.

4.4 MANAGING THE SPECIFICATIONS STAGE

For many inexperienced designers, the specification stage is the most difficult. It requires practice and effort to learn to think in terms of functions and to regard them as intellectual objects distinct from objectives and constraints. To this end, team-based approaches can be quite helpful.

There are some engineering tasks that are better done by individuals (e.g., calculating a stress in a bridge cable), while there are others that are best done by groups (e.g., designing a major suspension bridge). Setting specifications is a task that has both individual and group components. Many teams find it useful to have its members initially use the methods as individuals, and subsequently have the entire team review, discuss, and revise these individual results. This sort of divide-and-conquer approach ensures that team members are prepared for meetings and that there is an experienced practitioner for each tool or method. Group review and revision allows team members to build on each other's ideas in developing and setting functional specifications, so the team benefits from fresh viewpoints and further critical thinking.

4.5 NOTES

Section 4.1: Further details about the Quebec Bridge and other "engineer's dreams" can be found in Petroski's history of American bridge building (1995). The black box for the radio used in Section 4.1.1 was developed by our colleague, Carl Baumgaertner, for use in Mudd's freshman design course. The term glass box method was coined in (Jones, 1992). The function-means tree used was developed by our Harvey Mudd colleague, James Rosenberg, to illustrate an example originally proposed in (Akiyama, 1991).

Section 4.3: The results from the Xela-Aid chicken coop design project continue to be taken from final reports (Gutierrez et al., 1997) and (Connor 1997).

4.6 EXERCISES

4.1 Explain the differences between functions and objectives.

4.2 Explain the differences between metrics and performance specifications.

4.3 Using each of the methods for developing functions described in Section 4. 1, develop a list of the functions of the portable electric guitar of Exercise 3.2. How effective was each of these methods in developing the specific functions?

4.4 Using each of the methods for developing functions described in Section 4. 1, develop a list of the functions of the rainforest project of Exercise 3.5. How effective was each of these methods in developing the specific functions?

4.5 Based on the results of either Exercise 4.4 or 4.6, discuss the relationship between methods of determining functions, the nature of the functions being determined, and the nature of the artifact being designed. Is the designer's level of experience also likely to affect the outcome of functional analysis?

4.6 Do the research necessary to determine whether there are any applicable standards (e.g., safety standards, performance standards, interface standards) for the design of the portable electric guitar from Exercise 3.2.

4.7 Describe the interfaces between the portable electrical guitar of Exercise 3.2 and, respectively, the user and the environment. How do these interfaces constrain the design?

4.8 Developing countries often have different safety (and other) standards than are typically found in countries such as Canada and the United States. How could this affect the design of both the portable electric guitar of Exercise 3.2 and the rainforest project of Exercise 3.5?

4.9 What are the interface design boundaries and issues for the design and installation of a new toilet for a building?

CHAPTER 5

FINDING ANSWERS TO THE PROBLEM

How many good designs are there? Which one is best?

WE DESCRIBED the problem definition phase of the design process in Chapter 3 and design specifications in Chapter 4. Now we will explore the transition from conceptual design to preliminary design wherein we generate design concepts or schemes that will meet our objectives. Thus, we want to (1) generate potential designs, (2) organize them in ways that make their exploration easy, and (3) evaluate them to see which are worth pursuing and which are not. We will also discuss designing for quality and building proofs-of-concept and prototypes.

5.1 WHAT IS A DESIGN SPACE?

A *design space* is a mental construct of an intellectual space that envelops or incorporates all of the potential solutions to a design problem. As a broad concept, the utility of the notion of a design space is limited to its ability to convey a *feel* for the design problem at hand. The phrase *large design space* conveys an image of a design problem in which (1) the number of potential designs is very large, perhaps even infinite, or (2) the number of design variables is large, as is the number of values they can assume.

The "size" of a design space reflects the number of possible design solutions and the number of design variables.

Two designed artifacts that have large design spaces are passenger aircraft (e.g., the Boeing 747), and major office buildings (e.g., Chicago's Sears Tower)—engineered objects with a lot of parts. A 747 has six million different parts, and we can only imagine how many parts there are in a 100-story building, from window frames and structural rivets to water faucets and elevator buttons. With so many parts, there are still more design variables and design choices.

Thus, the 747 and the Sears Tower have very large design spaces. On the other hand, these artifacts differ from one another because their performance presents different challenges and airplanes are constrained in ways that buildings are not. The architect and structural engineer designing a high-rise building may have innumerably more choices for the shape, footprint, and structural configuration of a skyscraper than do aeronautical engineers who are designing fuselages and wings. While weight is important as the number of floors and occupants of a building mount, and while the shapes of high-rise buildings are analyzed and tested for their response to wind, they are subject to fewer constraints than are the payload and aerodynamic shape of aircraft. As we will discuss further in Section 5.2.3, constraints play an important role in limiting the size of a design space.

A *small* or *bounded design space* conveys the image of a design problem in which (1) the number of potential designs is limited or small, or (2) the number of design variables is small and they, in turn, can take on values only within limited ranges. The design of individual components or subsystems of large systems often occurs within small design spaces. For example, the design of windows in both aircraft and buildings is so constrained by opening sizes and materials that their design spaces are relatively small. Similarly, the range of framing patterns for low-rise industrial warehouse buildings is limited, as are the kinds of structural members and connections used to make up those structural frames.

5.1.1 Complex design spaces and decomposition

Large design spaces are complex because of the combinatorial possibilities that emerge when hundreds or thousands of design variables must be assigned. Design spaces are also complex because of interactions between subsystems and components, even when the number of choices is not overwhelming. In fact, one aspect of design complexity is that collaboration with many specialists is often critical because it is rare that a single engineer knows enough to make all of the design choices and analyses.

Divide and conquer, or decompose the problem.

Again, this complexity emerges because values of many design variables are highly dependent either on choices already made or on those yet to be made. How do we attack such complex design spaces? One approach is to apply the idea of *decomposition*, or *divide and conquer*, the process of dividing or breaking down a complex problem into subproblems that are more readily solved. Designs of airplanes, for example, can be decomposed into subproblems: the wings, the fuselage, the avionics, the tail, the kitchen, the passenger compartment, and so on. In other words, the design problem and the design space are broken into manageable pieces that are taken on one at a time!

The tools that we present in this chapter are designed to assist in decomposing a design problem into solvable subproblems and reassembling their solutions into coherent, feasible designs. The morphological chart described in Section 5.3 is particularly suited to (1) decomposing the overall functionality of a design into its constituent subfunctions, (2) identifying the means for achieving each of those functions, and (3) enabling the composition, or *recomposition*, of possible design solutions. The recomposition or *synthesis* of feasible or workable solutions is particularly important. Imagine how it would feel to take apart a very elaborate clock mechanism in order to fix something, and then finding that it could not be put back together because a new part was too large or a fitting had a slightly different configuration! Similarly, when we are recomposing candidate designs, we have to be sure to exclude incompatible alternatives, which is why we present in Section 5.4 some ideas and tools for ensuring that our newly assembled designs can work.

5.2 GENERATING DESIGN IDEAS: EXPANDING THE DESIGN SPACE

How do we generate design ideas? Or, how can we usefully expand the size and range of our design space? In this context, we note that:

- Engineering creativity is goal-directed and designed to serve a known purpose, not to search for one. The goal may be imposed externally, as is often the case in engineering design firms, or internally, as in a new product idea being developed in a garage. But, *there is a goal toward which the creative activity is aimed.*
- *Creative activity involves work* or, as Thomas Edison said, "Invention is 99 percent perspiration and one percent inspiration." In other words, we will likely fail if we are not prepared to do some serious work to generate design alternatives.

Therefore, in order to do good, goal-directed design generation, we ask, "What means are available to help us generate design ideas and to enlarge and organize the space of potential designs?"

Two central themes will emerge in our answers to these two questions. One is that we rarely gain any advantage by reinventing the wheel. In other words, and particularly when we define functions and the means to execute them, we should be aware that other people may have already tried to implement some of the functions we want to make happen. It is only common sense to suggest that we should identify, study, and then use the formidable amount of existing information already available. Thus, it should not be a surprise that much of what we will now discuss has parallels with our prior discussion of means for acquiring information (see Section 2.3.3.1).

The second principal theme is that the kinds and styles of team interactions that we describe below have much in common with brainstorming. That is, as we describe activities that a design team (perhaps including experts or other stakeholders) could undertake to generate design alternatives, we will also hark back to some of the ideas advanced in Section 3.5.1 about brainstorming. We will suggest both particular activities a group can undertake in Section 5.2.3 and ideas about how team members might think about things in Section 5.2.4.

5.2.1 Taking advantage of design information that is already available

In Section 2.3.3.1 we detailed the importance of conducting literature reviews to identify prior work in the field and determine the state of the art. This included locating and studying previous solutions, product advertising, vendor literature as well as handbooks, compendia of material properties, design and legal codes, etc. *The Thomas Register* is one valuable digest of product vendors. It lists more than one million manufacturers of the kinds of systems and components used in mechanical and electrical design. Annual updates of the 23-volume *Register* are found in most technical libraries. Further, while much more material is becoming available on the World Wide Web, far more information is not—and may never be. Thus, while web-surfing is a very useful form of information gathering, it should never be seen as the only way to identify and retrieve design-related knowledge.

Two information gathering tasks are central to product design. Competitive products are

- *benchmarked* to evaluate how *well* such existing products perform certain functions, and
- are *dissected* or *reverse engineered* to see *how* functions are performed and so identify other ways of performing similar functions.

We repeat these means here to drive home a point: Revisiting the notes taken while the original design goals were being clarified can help us generate design concepts and alternatives because it is likely that some ideas and solutions emerged during the problem definition phase. Thus, looking back at "old" notes allows us to recapture old or premature ideas that were recorded then.

5.2.2 Patents: Expanding a design space without reinventing the wheel

One related activity that we can undertake while generating alternatives is to search for relevant patents that have been awarded. We do this to avoid "reinventing the wheel" and to leverage our thinking by building on what we already know about a still-emerging design. We might also do a patent search to identify available technology that we can use in our design, assuming we can negotiate appropriate licensing agreements with the patent holder(s).

Patents are a kind of *intellectual property*, meaning that holders of patents are identified as those who are given the credit for having discovered or invented a device or a new way of doing things. A designer's list of patent awards carries a lot of weight in engineering practice. This intellectual credit is awarded by the U.S. Patent and Trademark Office (USPTO) to individuals and/or to corporations after they file an application that details what they believe to be the *new art* or originality of their invention or discovery. Two basic patents are awarded or denied by the USPTO after patent applications are thoroughly evaluated by USPTO patent examiners:

- *Design patents* are granted on the form or appearance or "look and feel" of an idea. They clearly relate to an object's visual appearance, as a result of which they are relatively weak patents because only minor alterations in the appearance of a device are enough to create a new product.
- *Utility patents* are granted for functions, that is, on how to do something or make something happen. They are stronger and often hard to "work around" because they focus on function rather than form.

In either case, it should be kept in mind that a patent reflects the fact that someone, the patent holder, has been identified as the *owner* of this intellectual property.

A computer-based version of an index to patents, *Classification and Search Support Information System* (CASSIS), can be found at most libraries. Patents are listed by individual class and subclass numbers that are detailed in a rather complex classification index. The USPTO maintains its own Web site (and search engine) that presents data about granted patents, and it publishes a weekly edition of the *Official Gazette* that lists, in numerical order by class number, all patents granted in the previous week.

Since patents result from (examined) *claims*, they can be—and often are—challenged because others feel that they have developed the relevant *prior art* that forms the foundation of that challenged patent. Filing patent applications is a mixed blessing for the designer. The award of a patent provides protection for new and innovative ideas. At the same time, however, patents may inhibit the development of ideas for second- and third-generation improvements of existing devices or processes.

5.2.3 Group activities for the design team

Brainstorming was earlier identified as a small-group activity done to clarify a client's goals. We pointed out that new ideas, some related, some not, might be generated, and we suggested that these new ideas be recorded but not evaluated. We also suggested that design team members show respect for the ideas and offerings of their teammates. We now extend those behavioral themes in the context of design generation, emphasizing some "rules of the game" for three different "games" that teams can "play" to generate design alternatives.

Successful design calls for two different kinds of thought, divergent thinking and convergent thinking:

- *Divergent thinking* is done when we try to remove limits or barriers, hoping instead to be expansive while trying to increase our store of design ideas and choices. Thus, we want to "think outside of the box" or "stretch the envelope" when trying to expand the design space and generate design alternatives.

- *Convergent thinking* describes what we do to narrow the design space to focus on the "best" alternative(s) after we have opened up the design space sufficiently. Our problem solving takes on a narrow focus so as to converge to a solution within known boundaries or limits.

Design, a goal-oriented activity, combines divergent and convergent thinking. The design process seeks to converge upon the best possible solution, however "best" is defined. At the same time, the process' activities that lead us to converge on a solution require divergent thinking. For example, we want to open up the problem space to better understand the problem and identify other stakeholders. Similarly, we engage in divergent thinking when we expand our functions to cover the most general case. After completing such "divergent" activities we try to converge to an adequate representation of our understanding. Thus, setting and documenting priorities among objectives is a convergent process, as is selecting appropriate metrics for our objectives. This interplay of modes of thinking (and acting) is part of what gives design its intellectual richness.

Think outside the box, but within the physics!

We now describe three intuitive activities that work to encourage divergent "free thinking" to enhance our collective creativity, although we do *not* mean to encourage thinking that violates the axioms of logic or physical principles. People working in groups may interact more spontaneously, in a free-spirited way that calls forth associations from group members in ways that we cannot anticipate or logically force. These activities are also *progressive* in nature because there are iterative cycles within each that results in the progressive emergence and refinement of new design ideas.

5.2.3.1 The 6–3–5 method

The first group design activity is the *6–3–5 method*. The name derives from having *six* team members seat themselves around a table to participate in this idea generation game, each of whom writes an initial list of *three* design ideas briefly expressed in key words and phrases. The six individual lists are then circulated past each of the remaining group members in a sequence of *five* rotations of *written* comment and annotation. Verbal communication or cross-talk is not allowed. Thus, each list makes a complete circuit around the table, and each member of the group is stimulated in turn by the increasingly annotated lists of the other team members. When all of the participants have commented on each of the lists, the team would use a common visualization medium (e.g., a blackboard) to list, discuss, evaluate, and record all of the design ideas that have resulted from a group enhancement of all of the team members' individual ideas.

We can generalize this method to the "$m - 3 - (m - 1)$" method by starting with m team members and using $m - 1$ rotations to complete a cycle. However, the logistics of ever-lengthening lists written on increasingly-crowded sheets of paper, and of providing tables that seat more than six, suggest that six may be a "natural" upper limit for this activity. And especially in an academic setting, we would prefer fewer than six—ideally no more than four—on a project team.

You don't have to be an artist to be a visual thinker.

5.2.3.2 The C-sketch method

The *C-sketch method* starts with an initial sketch of a single design concept by each of the team members, and then proceeds as does the 6–3–5 method. Each sketch is circulated through the team in the same fashion as the lists of ideas in the 6–3–5 method, with all of the annotations or proposed design modifications being written or sketched on the initial concept sketches. Again, the only permissible form of communication is by pencil on paper, and the discussions that follow the end of a complete cycle of sketching and modifying follow those described in the 6–3–5 method. Research suggests that the C-sketch method becomes unwieldy with even five team members due to the crowding of annotations and modifications on a given sketch. However, the C-sketch method is very appealing in an area such as mechanical design because there is strongly suggestive evidence that sketching is a natural form of thinking in mechanical device design. Research has also shown that drawings and diagrams facilitate the grouping of relevant information added in marginal notes, and they help people to better visualize the objects being discussed.

5.2.3.3 The gallery method

The *gallery method* is a third approach to getting team reactions to drawings and sketches, although the sketching and communication cycles are handled differently. In the gallery method, group members first develop their individual, initial ideas within some allotted time, after which all of the resulting sketches are posted, say on a corkboard or a conference room whiteboard. This set of sketches forms the backdrop for an open, group discussion of *all* of the posted ideas. Questions are asked, critiques offered, and suggestions made. Then each participant returns to her or his drawing and suitably modifies or revises it, again within a specified period of time, with the goal of producing a second generation idea. The gallery method is thus both iterative and progressive, and there is no way to predict just how many cycles of individual idea generation and group discussion should be held. Our only recourse would be to apply the common sense dictum of the *law of diminishing returns*: We proceed until a consensus emerges within the group that one more cycle will not produce much (or any) new information, at which point we quit.

5.2.4   Ways to think divergently

A *metaphor* is a figure of speech. It is a style in which attributes of one object or process are used to give depth or color to the description of a second object or process. For example, to describe engineering education as drinking from a fire hose is to suggest that engineering students are expected to absorb a great deal of knowledge rapidly and under great pressure. We use metaphors to point out *analogies* between two different situations, that is, to suggest that there are parallels or similarities in the two sets of circumstances. Analogies can be very powerful tools in engineering design. One of the most often cited is the Velcro fastener, for which the analogy drawn was to those pesky little plant burrs that seem to stick to everything on which they're blown.

The Velcro fastener resulted from a *direct analogy* in which its inventor made a direct connection between the individual elements of the plant burr and the connecting fibres of the fastener. We also can use *symbolic analogies*, as when we plant ideas or talk about objectives trees. In these cases we are clearly drawing connections through some underlying symbolism. We might also apply *personal analogies* by imagining what it would feel like to be the object (or part of it) that we're trying to design. For example, how might it feel to be a tin beverage container with a pull-tab? We could also stray into the realm of *fantasy analogies* by imagining something that is literally fantastic or beyond belief. Of course, looking at our world at the end of the second millennium, just how much more fantastic can it get than space travel, instantaneous and reliable personal communication across the globe, and seeing clearly into the human body with CT scans and magnetic resonance imaging (MRIs).

The impossible is often the starting point for the excellent!

Fantasy analogies suggest another approach to "thinking out of the box." We are not very far past the time when many of the technologies we take for granted were thought to be outrageous ideas that were beyond belief. When Jules Verne published his classic *20,000 Leagues Under the Sea* in 1871, the idea of ships that could "sail" deep in the ocean was viewed as an outrageous fantasy. Now, of course, submarines and seeing unfamiliar yet exciting life forms under water are part of everyday experience. We cannot escape the idea that design teams might imagine the most outrageous solutions to a design problem and then seek ways to make such solutions useful. For example, airplanes that are invisible to radar were once considered far fetched. The arterial stents used in angioplastic surgery are also devices once thought to be impossible (see Figure 5.1). Who would have believed that an engineering structure could be erected within the narrow (3–5 mm in diameter) confines of a human artery?

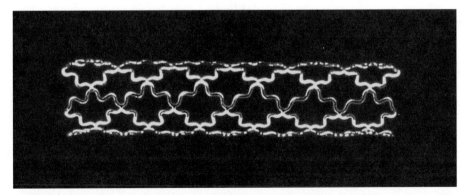

FIGURE 5.1   This is a PALMAZ-SCHATZ™ balloon expandable coronary *stent*, a device used to maintain arterial shape and size so as to allow uninhibited and natural blood flow. Note how this structure resembles the kind of scaffolding often seen around building renovation and construction projects. Photo courtesy of Cordis, a Johnson & Johnson Company.

The stent suggests still another aspect of analogical thinking, namely, looking for *similar solutions*. The stent clearly is similar in both intent and function to the scaffolding erected to support walls in mines and tunnels as they are being built. Thus, the stent and the scaffold are *like ideas*.

We could invert this idea by looking for *contrasting solutions* in which the conditions are so different, so contrasting, that a transfer of solutions would seem totally implausible. Here we would be looking for *opposite ideas*. Fairly obvious contrasts would be between strong and weak, light and dark, hot and cold, high and low, and so on. One example of using an opposite idea occurs in guitar design. Most guitars have their tuning pegs arrayed at the end of the neck. In order to make a portable guitar, one clever designer chose to put the tuning pegs at the other end of the strings, at the bottom of the body, in order to save space and thus accentuate the guitar's portability.

Finally, in addition to finding similar and contrasting solutions, we recognize a third category. *Contiguous solutions* are developed by thinking of *adjoining* (or *adjacent*) *ideas* in which we take advantage of natural connections between ideas, concepts and artifacts. For example, chairs prompt us to think of tables, tires prompt us to think of cars, and so on. Contiguous solutions are distinguished from similar solutions by their adjacency, that is, bolts are adjacent to nuts and are contiguous solutions, while bolts and rivets serve identical fastening functions and are thus similar solutions.

5.3   LIMITING THE DESIGN SPACE TO A USEFUL SIZE

On the more general subject of guiding the search for design solutions, there are some pragmatic issues relating to how an artifact will be both used and made that provide guideposts for developing a search space. However, the guideposts may narrow, rather than expand the search space. In particular, the guideposts suggest that design alternatives be analyzed as "functions" of:

- user needs,
- available technologies, and
- external constraints.

For example, the design of vehicles for a campus transportation system will be influenced by the *user needs* of potential users. Candidate vehicles might include low-end, simple bikes; nifty, high-tech bikes; recumbent bikes; tricycles; or even rickshaws. Among user needs that might affect the consideration and design of these vehicles would be the availability of parking facilities at lecture and residence halls, the need to carry packages, the need for access for the handicapped, and so on. User needs could have a major impact on the kinds of features called for in a campus vehicle, and thus on the very vehicle type. Just as clearly, the *availability of different technologies* might stimulate the process of generating alternatives. Imagine the different materials of which a bike or bike-like device could be made: The choice of materials affects a bike's appearance, its manufacture, and its price. Finally, *external constraints* may affect the design. Examples of such constraints include a team's competence in certain design fields (e.g., it may be more comfortable designing tricycles than high-tech performance bikes) or the availability of manufacturing capabilities (e.g., it should avoid designing a bike made of composite materials if the only available manufacturing facility forms and connects metals).

Similarly, there are practical considerations to keep in mind to ensure that the size and scope of a design space are manageable. Such considerations are often issues of *common sense*. In particular, lest the lists of issues and of candidate designs become too large or too silly, we can:

- *Invoke and apply constraints,* in much the same way we did above while assessing the influence and importance of user needs.
- *Freeze the number of attributes* being considered to avoid details that are unlikely to seriously affect the design at this point (e.g., the color of a bike or car is not worth noting in a design's early stages).
- *Impose some order on the list,* perhaps by harking back to data gathered during problem definition that suggests that particular functions or features are more important.
- *"Get real!"* or, in other words, watch out when silly, infeasible ideas are repeated often. Common sense, however, must be applied in ways consistent with our earlier admonitions about maintaining supportive environments for design brainstorming.

We will have more to say about managing the process of generating and selecting design alternatives in Section 5.7.

5.4 MORPHOLOGICAL CHARTS: ORGANIZING FUNCTIONS AND MEANS TO GENERATE DESIGNS THAT WORK

Morphological charts provide another way to visualize elements of a design space. In fact, *morph charts,* as they are affectionately called, show the design space in a way that gives us a sense of its size while allowing us to identify potential designs.

We begin by constructing a list of either features that we want the design to have or functions we want it to perform. The list should be of a reasonable and manageable size, with all features or functions being at the same level of detail. Say within their corresponding objectives or function-means trees, because it helps ensure internal consistency. We then list all of the different means of implementing each function or feature identified. Thus, for example, a list of features and functions for the beverage container problem consistent with the subobjectives of the objective Promote Sales (see Figure 3.2) might look like:

Contain Beverage

Material for Beverage Container

Provide Access to Juice

Display Product Information

Sequence Manufacture of Juice and Container

We then build lists of means for achieving each of these functions and attributes and attach them to their corresponding entries, thus creating a table:

Contain Beverage:	Can, Bottle, Bag, Box
Material for Beverage Container:	Aluminum, Plastic, Glass, Waxed Cardboard, Lined Cardboard, Mylar Films
Provide Access to Juice:	Pull-Tab, Inserted Straw, Twist-Top, Tear Corner, Unfold Container, Zipper
Display Product Information:	Shape of Container, Labels, Color of Material
Sequence Manufacture of Juice and Container:	Concurrent, Serial

This tabulated information provides what we need to construct a morph chart. In fact, we show the resulting *morphological chart* in Figure 5.2, where the same information is displayed in a visually appealing and usefully organized form. The features and functions that the device must serve are listed on the vertical axis, while for each of these attributes two or more means are identified and listed in cells in that row. A conceptual design can be constructed by linking one means, any means, for each identified function, subject only to interface constraints that may prevent a particular combination. For example, one design could consist of a plastic bottle with a twist top, its color chosen to correspond to a particular beverage, and made (and stored) in advance of the beverage's expected delivery. We are thus assembling designs in the classic "Chinese menu" style, choosing one means from

MEANS FEATURE/FUNCTION	1	2	3	4	5	6
Contain beverage	can	bottle	bag	box	••••	••••
Material for drink container	aluminum	plastic	glass	waxed cardboard	lined cardboard	mylar films
Mechanism to provide access to juice	pull tab	inserted straw	twist top	tear corner	unfold container	••••
Display of product information	shape of container	labels	color of material	••••	••••	••••
Sequence manufacture of juice container	concurrent	serial	••••	••••	••••	••••

FIGURE 5.2 A morphological chart for the beverage container design problem. The functions that the device must serve are listed on the vertical axis, while for each of them two or more means are identified. Subject only to interface constraints that may prevent a particular combination, a conceptual design or scheme can be constructed by linking one means, any means, for each of the five identified functions, thus assembling a design in the classic "Chinese menu" style.

each of rows *A*, *B*, *C* . . . to combine into a design scheme. Similarly, the morph chart is analogous to a spreadsheet that similarly enables certain kinds of "calculations."

How many potential solutions are identified in a morph chart? Or, in other words, just how big is our design space? The correct answer would account for the *combinatorics* that result from combining any single means in a given row with each of the remaining means in all of the other rows. Thus, for the beverage container morph chart of Figure 5.2, the number of candidate designs could be as large as $4 \times 6 \times 6 \times 3 \times 2 = 864$.

Morph charts include all of the alternatives; infeasible ones must be pruned.

However, having done the calculation and seen how large a design space can become, it is important to recognize that not all of these 864 connections are, in fact, feasible or candidate designs. For example, we are unlikely to design a glass bag with a zipper! Thus, while building a morphological chart provides a way to expand our design space and identify alternatives, it also provides the opportunity to *prune the design space by identifying and excluding incompatible alternatives*. We cannot assume that all of the connections made by following out the combinatorial arithmetic are valid connections. There are clearly other

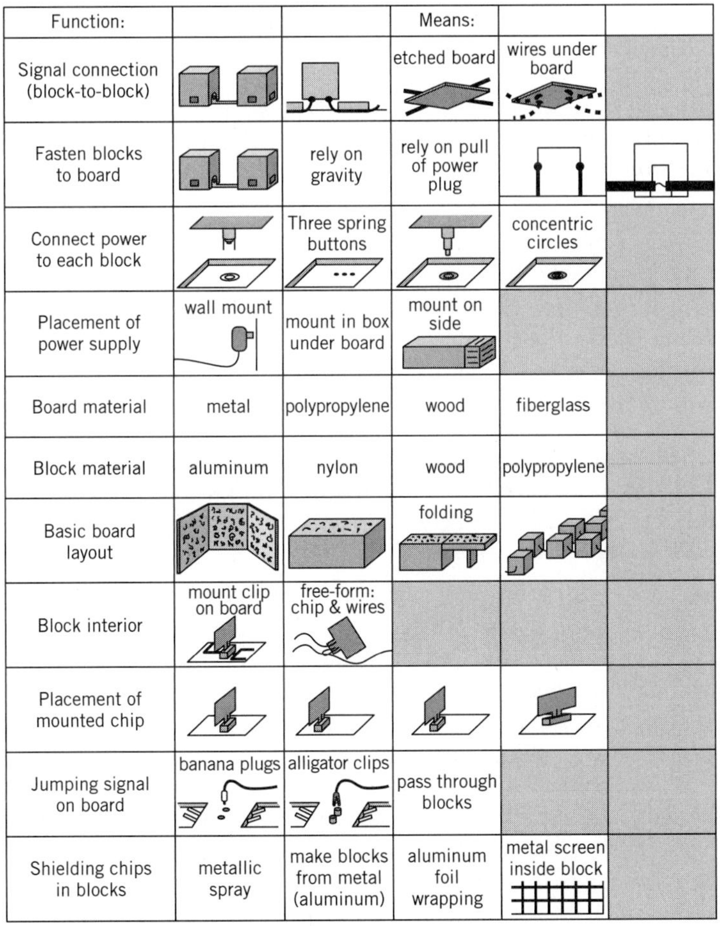

FIGURE 5.3

A morph chart for a "building block" analog computer that was done in Harvey Mudd's E4 design course (Hartmann, Hulse et al. 1993).

combinations of container designs that cannot work (e.g., glass cans with pull-tabs or inserted straws) and therefore must be excluded from the design space. To exclude such alternatives we can apply design constraints, physical principles, and plain common sense. We should also remember that technologies and, consequently, available means, do change over time. For example, the waxed paper container has evolved from being one that only supported the containment function, to a modern one that incorporates a twist-cap at its top, so that this container supports both containment and intermittent mixing (after it has been opened) of the contents, which is why orange juice cartons have such tops and milk cartons don't.

It is important to list features and functions at the same level of detail when building a morph chart because we don't want to compare apples to oranges. Thus, for the beverage container, we would not include means of Resisting Temperature and means for Resisting Forces and Shocks within the morph chart of Figure 5.2 because they are more detailed functions that derive from subgoals that are much further down in the objectives tree of Figure 3.2. Similarly, when doing a complex design task, such as designing a building, we don't want to worry about means for identifying exits or for opening doors while developing different concepts for moving between floors, which might include elevators, escalators, and stairways.

As a further illustration, we show in Figure 5.3 a morph chart constructed in a freshman design project done at Harvey Mudd, this one being concerned with the design of a "building block" analog computer. This morphological chart is appealing in its use of graphics and icons to illustrate many of the means that could be applied to achieve its functions. This style of morph chart may also be useful in the kind of C-sketch discussions described in Section 5.2.3.2.

Similarly, morph charts can also be used to expand the design space for large, complex systems by listing principal subsystems in a starting column and then identifying various means of implementing each of the subsystems. For example, to design a vehicle, one subsystem would be its Source of Power, for which the corresponding means could be Gasoline, Diesel, Battery, Steam, and LNG. Each of these power sources is itself a subsystem that needs further detailed design, but the array of power subsystems expands the range of our design choices.

5.5 CONNECTING CONCEPTS TO OBJECTIVES: SELECTING THE BEST DESIGNS

A good job of generating design concepts will likely produce several schemes or alternatives from which to choose. We may have used a morph chart to identify a number of design schemes, or we may have generated alternatives with a less structured approach. But no matter how it was done, we have to "pick a winner" among the identified options and select one or (perhaps) two concepts for further elaboration, testing, and evaluation. We choose only one or two simply because time, money, and personnel resources are always limited.

Many approaches are used to assess design alternatives. Some are formal and seemingly rigorous. Others are simple, such as picking the one "liked" the best, having a client or customer make a choice for reasons unspecified, or having an executive or a champion make a decision based on personal criteria.

We will use three methods to select from among a set of alternative designs or concepts, each a variant of the *Pugh selection chart*: the numerical evaluation matrix, the weighted checkmarks chart, and the best of class chart. Further, we will cast the methods in terms of the design *objectives* because designs should meet the objectives of all of the stakeholders, including clients, users, the public at large, and so on.

In addition, the objectives we use are weighted, a calculation that has some clear limits. As noted in Section 3.3.4, we can weight objectives using a PCC, but *only if those objectives stem from a common node*. Thus, the weightings that appear in the discussions below were so constrained and determined.

The three methods we suggest for selecting a "best" design explicitly link design alternatives to objectives but do not have the mathematical rigor that goes along with finding maxima and minima in calculus. Rather, we are striving to bring some order to judgments and assessments that are subjective at their root. Just as professors give grades to encapsulate judgments about how well students have mastered concepts, ideas, or methods, designers try to integrate the best judgments of stakeholders and design team members in ways that allow these judgments to be exploited in a sensible and orderly manner. We must use common sense when we look at the results of the methods we now describe.

No matter which of the charts we apply, our first step should always be to evaluate each alternative in terms of all of the constraints that apply since alternative designs must be rejected if the constraints are not met. As we outline our three selection methods, we will assume that all applicable constraints are being applied and that the design space is being narrowed accordingly.

5.5.1 Numerical evaluation matrices

We show *numerical evaluation matrices* for the BJIC and GRAFT companies in Figures 5.4 and 5.5. These charts show objectives and calculated weights in the left-hand columns, while the scores assigned to each objective are shown in design-specific columns on the right. The constraints for this beverage container problem are shown at the top of the charts, and by applying them we would rule out glass bottles and aluminum containers because of potential sharp edges. This reduces the number of designs to two, the Mylar bag and the polyethylene bottle. We now score these two alternatives against the objectives detailed in Chapter 3 and weighted as described above.

We see that the polyethylene bottle scores 0.9 for environmentally benign, while the Mylar bag earns a score of 0.1 for this metric. (Note that all of the metrics results have been normalized to a 0–1 range.) Since the BJIC Company values environmentally benign much more (i.e., 33 percent) than the GRAFT Company (i.e., 4 percent), the scores earned for each candidate design for this metric are significantly lower for BJIC than for GRAFT. The cumulative results for all of the objectives for the two remaining viable products show that BJIC's values rate the polyethylene bottle significantly ahead, while GRAFT's values dictate a choice of the Mylar bag by a similar margin.

Beyond these calculated results about these two hypothetical candidate designs, the most important feature in Figures 5.4 and 5.5 is that each chart has the same value for the metrics applied to the design alternatives. Recall from our discussion in Section 4.3 that metrics are measurable indicators of how well specific objectives are met. Thus, if our metrics had different values for different design alternatives, we would have to assume that the testing process is defective. Here the design team has clearly selected different alternatives as a proper reflection of the fact that their different clients (i.e., BJIC and GRAFT) have

DESIGN Constraints and Objectives	Weight (%)	Glass bottle with twist-off cap	Aluminum can with pull-tab	Polyethylene bottle with twist-off cap	Mylar bag with straw
C: No sharp edges		✘	✘		
C: No toxin release					
C: Preserves quality					
O: Environmentally benign	33			0.9 × 33% 29.7%	0.1 × 33% 3.3%
O: Easy to distribute	09			0.5 × 9% 4.5%	0.6 × 9% 5.4%
O: Preserves taste	22			0.9 × 22% 19.8%	1.0 × 22% 22%
O: Appeals to parents	18			0.8 × 18% 14.4%	0.5 × 18% 9.0%
O: Permits marketing flexibility	04			0.5 × 4% 2.0%	0.5 × 4% 2.0%
O: Generates brand identity	13			0.2 × 13% 2.6%	1.0 × 13% 13%
TOTALS	99			73.0%	54.7%

FIGURE 5.4 A numerical evaluation matrix for the beverage container design problem. This chart reflects BJIC's values in terms of the weights assigned to each objective, as given in the pairwise comparison chart of Figure 3.4 (b).

weighted their objectives differently, perhaps because they have different corporate values. There was no difference in the design selection process in either the metrics or the testing procedures.

It is worth noting that this might not be the case if the companies were independently doing their designs and, consequently, rating each product on its different dimensions. That is, it is not at all hard to imagine that some companies might find a Mylar bag significantly more expensive to produce and distribute than they would a polyethylene bottle. In such a case, the metric for low cost of production and distribution might be lowered from 0.6 to 0.1, for example, from which a rather different outcome would emerge.

5.5.2 The weighted checkmark method

The *weighted checkmark method* is a simpler, qualitative version of a numerical evaluation matrix. We simply rank the objectives as high, medium, or low in priority. Objectives with

DESIGN Constraints and Objectives	Weight (%)	Glass bottle with twist-off cap	Aluminum can with pull-tab	Polyethylene bottle with twist-off cap	Mylar bag with straw
C: No sharp edges		✖	✖		
C: No toxin release					
C: Preserves quality					
O: Environmentally benign	04			$0.9 \times 4\%$ 3.6%	$0.1 \times 4\%$ 0.4%
O: Easy to distribute	22			$0.5 \times 22\%$ 11.0	$0.6 \times 22\%$ 13.2%
O: Preserves taste	09			$0.9 \times 9\%$ 8.1	$1.0 \times 9\%$ 9%
O: Appeals to parents	13			$0.8 \times 13\%$ 10.4%	$0.5 \times 13\%$ 6.5%
O: Permits marketing flexibility	18			$0.5 \times 18\%$ 9.0%	$0.5 \times 18\%$ 9.0%
O: Generates brand identity	13			$0.2 \times 33\%$ 6.6%	$1.0 \times 33\%$ 33%
TOTALS	99			48.7%	74.7%

FIGURE 5.5 A numerical evaluation matrix for the beverage container design problem. This chart reflects GRAFT's values in terms of the weights assigned to each objective, as shown in the pairwise comparison chart of Figure 3.4 (a). However, note that the scores found for each metric in the chart are the same as those used for the BJIC design and shown in Figure 5.4. Is that as it should be? If so, why?

high priority are given three checks, those with medium priority are given two checks, while objectives with low priority are given only one check, as shown in Figure 5.6. Similarly, metrics are taken as 1 if rated greater than 0.5, and as 0 if their rating is less than 0.5. If a design alternative meets an objective in a "satisfactory" way, it is then marked with one or more checks, as shown in Figure 5.6. Finally, the number of checks are summed over all of the valid alternatives, the constraints having already been applied. This method is easy to use, makes the setting of priorities rather simple, and is readily understood by clients and other parties. On the other hand, the weighted checkmark approach lacks detailed definition and it sets up all of the metrics as binary variables that are either checked (satisfactory) or not. This makes it easy to succumb to temptation and "cook the results" to achieve a desired outcome.

DESIGN Constraints and Objectives	Weight (%)	Glass bottle with twist-off cap	Aluminum can with pull-tab	Polyethylene bottle with twist-off cap	Mylar bag with straw
C: No sharp edges		✖	✖		
C: No toxin release					
C: Preserves quality					
O: Environmentally benign	✓✓✓			1 × ✓✓✓ ✓✓✓	0 × ✓✓✓ ••••
O: Easy to distribute	✓			1 × ✓ ✓	1 × ✓ ✓
O: Preserves taste	✓✓			1 × ✓✓ ✓✓	1 × ✓✓ ✓✓
O: Appeals to parents	✓✓			1 × ✓✓ ✓✓	1 × ✓✓ ✓✓
O: Permits marketing flexibility	✓			1 × ✓ ✓	1 × ✓ ✓
O: Generates brand identity	✓✓			0 × ✓✓ ••••	1 × ✓✓ ✓✓
TOTALS				9✓	8✓

FIGURE 5.6   A weighted benchmark chart for the beverage container design problem. This chart qualitatively reflects BJIC's values in terms of the weights assigned to each objective, so it is a qualitative version of the evaluation matrix of Figure 3.4.

5.5.3 The best of class chart

Our last method for evaluating and ranking alternatives is the *best of class chart*. For each objective, we assign increasing scores to each design alternative that range from 1 for the alternative that meets that objective best, 2 to second-best, and so on, until the alternative that met the objective worst is given a score equal to the number of alternatives being considered. If, for example, there are 5 alternatives, then the best at meeting a particular objective would receive a 1, the second-best a 2, etc. Ties are allowed (e.g., two alternatives are considered "best" and so are tied for first) and are handled by splitting the available rankings (e.g., the two "firsts" would each get a score of $(1 + 2)/2 = 1.5$). These scores are then weighted in accord with the weighted objectives and the rest of the calculation would proceed as in Figures 5.4 and 5.5. The *lowest* summed score would be considered to be the best alternative design under this scheme.

    The best of class approach has its advantages and its disadvantages. One advantage is that it allows us to rank alternatives with respect to a metric, rather than simply treat as a binary, "yes or no" decision, as we did with weighted benchmarks. It, too, is relatively easy to implement and explain, and it can be done by individual team members or by a team as a group to make explicit any differences in rankings or approaches. The disadvantages of this approach are that it encourages evaluation based on opinion rather than testing or actual metrics, and it may lead to a moral hazard similar to that attached to weighted checkmarks, that is, the temptation to fudge the results or cook the books.

There is no excuse for accepting results blindly and uncritically.

No matter which of the three selection methods is used, there is no excuse for accepting the results blindly and uncritically. First of all, common sense must be applied when evaluating results. If two alternatives have relatively close scores, they should be treated as tied, unless there are unevaluated strengths or weaknesses or untested metrics. Second, if the evaluation results are unexpected, we should ask whether our expectations were simply wrong, whether the method was consistently applied, or whether our weights are not really appropriate to the problem. Third, if the results meet our expectations, we should ask whether they represent a fair application of the evaluation process, or have we just reinforced preconceived ideas or biases? Finally, if some alternatives have been rejected because they violated constraints, it might be wise to ask whether those constraints are truly binding.

5.5.4   Concept screening

It is also worth mentioning that the relative ease of using the foregoing methods, especially the weighted benchmark method, suggests that they might also be used to do informal *concept screening*, perhaps earlier in the design process, as a way to easily narrow the field of candidate designs. We could further facilitate a rapid screening with the weighted benchmark method, and the other selection tools, by taking the weights out of the matrices or calculations. Such screening could also be made a group process by clustering the checkmarks or some other symbols (e.g., dots) according to the number of people who vote for a concept. In this case, the number of votes would be easily evident from a visual display of the group members' votes.

5.6   PROTOTYPES, MODELS, AND PROOFS OF CONCEPT

We now discuss three-dimensional physical realizations of concepts for designed artifacts; that is, we will talk about objects made to strongly resemble the object being designed, if not actually mimic "the real thing." There are several versions of physical things that could be made, including prototypes, models, and proofs of concept, and they are all often made by the designer.

Prototypes are "original models on which something is patterned." They are also defined as the "first full-scale and usually functional forms of a new type or design of a construction (as an airplane)." In this context, prototypes are working models of designed artifacts. They are tested in the same operating environments in which they're expected to function as final products. It is interesting that aircraft companies routinely build prototypes, while rarely, if ever, does anyone build a prototype of a building.

A *model* is "a miniature representation of something," or a "pattern of something to be made," or "an example for imitation or emulation." We use models to *represent* some devices or processes. They may be paper models or computer models or physical models. We use them to illustrate certain behaviors or phenomena as we try to verify the validity of an underlying (predictive) theory. Models are usually smaller and made of different materials than are the artifacts they represent, and are typically tested in a laboratory or in some other controlled environment to validate their expected behavior.

5.6.1 Are prototypes and models the same thing?

A prototype is **the first of its kind**; a model **represents** a device or a process.

The definitions of prototypes and models sound enough alike that it prompts a question: Are prototypes and models the same thing? The answer is, "Not exactly." The distinctions between prototypes and models may have more to do with the intent behind their making and the environments in which they are tested than with any clear dictionary-type differences. Prototypes are intended to demonstrate that a product will function as designed, so they are tested in their actual operating environments or in similar, uncontrolled environments that are as close to their relevant "real worlds" as possible. Models are intentionally tested in controlled environments that allow the model builder (and the designer, if they are not the same person) to understand the particular behavior or phenomenon that is being modeled. An airplane prototype is made of the same materials and has the same size, shape, and configuration as those intended to fly in that series (i.e., Boeing 747s or Airbus 310s). A model airplane would likely be much smaller. It might be "flown" in a wind tunnel or for sheer enjoyment, but it is not a prototype.

Engineers often build models of buildings, for example, to do wind-tunnel testing of proposed skyscrapers, but these are not prototypes. Rather, building models used in a wind-tunnel simulation of a cityscape with a new high-rise are essentially toy building blocks meant to imitate the skyline. They are not buildings that work in the sense of aircraft prototypes that actually fly. So, why do aeronautical engineers build prototype airplanes, while civil engineers do not build prototype buildings? What do they do in other fields?

5.6.2 When do we build a prototype?

The answer is, "It depends." The decision to build a prototype depends on a number of things, including: the size and type of the design space, the costs of building a prototype, the ease of building that prototype, the role that a full-size prototype might play in ensuring the widespread acceptance of a new design, and the number of copies of the final artifact that are expected to be made or built. Aircraft and buildings provide interesting illustrations because of ample commonalities and sharp differences. The design spaces of both aircraft and high-rises are large and complex. There are, literally, millions of parts in each, so many, many design choices are made along the way. The costs of building both airplanes and tall buildings are also quite high. In addition, at this point in time, we have ample experience with both aeronautical and structural technologies, so that we generally have a pretty good idea of these two domains. So, again, why prototype aircraft and not prototype buildings? In fact, doesn't the complexity and expense of building a prototype aircraft argue directly against the idea of building such a prototype?

Notwithstanding all of our past experience with successful aircraft, we build prototypes of airplanes because, in large part, the chances of a catastrophic failure of a "paper design" are still unacceptably high, especially for the highly regulated and very competitive commercial airline industry that is the customer for new civilian aircraft. In other words, we are simply not willing to pay the price of having a brand new airplane take off for the first time with a full load of passengers, only to watch hundreds of lives being lost—as well as the concomitant loss of investment and confidence in future variants of that particular plane. It is an ethical issue because we bear responsibilities for technical decisions when they impinge upon our fellow humans. It is also an economic issue because the cost of a prototype is economically justifiable when weighed against potential losses. Also, we build prototypes of airplanes because they are not simply thrown away as "losses" after testing; prototypes are retained and used as the first in the series of full-size designs.

Buildings do fail catastrophically, during and after construction. However, they occur so rarely that there is little perceived value in requiring prototype testing of buildings before occupancy. Building failures are rare in part because high-rises can be tested, inspected, and experienced gradually, as they are being built, floor by floor. The continuous inspection that takes place during the construction of a building has its counterpart in the numerous inspections and certifications that accompany the manufacture and assembly of a commercial airliner. But the maiden flight of an airplane is a binary issue—the plane either flies or it doesn't—and a failure is not likely to be a graceful degradation!

Another interesting aspect of comparing the design and testing of airplanes to that of buildings pertains to the number of copies being made. We have already noted that prototype aircraft are not discarded after their initial test flights; they are flown and used. In fact, airframe manufacturers are in business to build and sell as many copies of their prototype aircraft as they can, so engineering economics plays a role in the decision to build a prototype. The economics are complicated because the manufacturing cost of the first plane in a series is very high. Technical decisions are made about the kinds of tooling and the numbers of machines needed to make an airplane, and economic trade-offs between the anticipated revenue from the sale of the airplane and the cost the manufacturing process are evaluated. We will address some manufacturing cost issues in Chapter 8.

Another lesson we can learn from thinking about buildings and airplanes is that there is no obvious correlation between the size and cost of prototyping—or the decision to build a prototype—and the size and type of the design space. And while it might seem that the decision to build a prototype may be strongly influenced by the relative ease of building it, the aircraft case shows that there are times when costly, complicated prototypes must be built. On the other hand, if it is cheap and easy to do, then it generally would seem a good idea to build a prototype. There certainly are instances where prototypes are commonplace, for example, in the software business. Long before a new program is shrink-wrapped and shipped, it is alpha- and beta-tested as early versions are prototyped, tested, evaluated, and, hopefully, fixed.

If there is a single lesson about prototypes, beyond that it is generally good to build them, it is that the project schedule and budget should reflect plans for doing so. More often than not a prototype is required, although there may be instances in which resources or time are not available. In weapons development contracts, for example, the U.S. Department of Defense virtually always requires that design concepts are demonstrated so that their performance can be evaluated before costly procurements. At the same time, it is interesting that aircraft companies (and others) are demonstrating that advances in computer-aided design and analysis allow them to replace some elements of prototype development with sophisticated simulation.

Sometimes we build prototypes of parts of large, complex systems to use as models to check how well those parts behave or function. For example, structural engineers build full-size connections, say, at a point where several columns and beams intersect in a geometrically complicated way, and test them in the laboratory. Similarly, aeronautical engineers build full-size airplane wings and load them with sandbags to validate analytical models of how these wing structures behave when loaded. A prototype of a part of a larger artifact was built in both instances, and then used to model behavior that needed to be understood as part of completing the overall design. Thus, again, we use prototypes to demonstrate functionality in the real world of the object being designed, and we use models in the laboratory to investigate and validate behavior of a miniature or of part of a large system.

5.6.3   Testing prototypes, models, and concepts

We introduced testing in discussing both models and prototypes. In design the type of testing that is often most important is *proof of concept* testing in which a new concept, or a particular device or configuration, can be shown to work in the manner in which it was designed. When Alexander Graham Bell successfully summoned his assistant from another room with his new-fangled gadget, Bell had proven the concept of the telephone. Similarly, when John Bardeen, Walter Houser Brattain, and William Bradford Shockley successfully controlled the flow of electrons through crystals, they proved the concept of the solid-state electronic valve, know as the transistor, that replaced vacuum tubes. Laboratory demonstrations of wing structures and building connections can also be considered as proof-of-concept tests when they are used to validate a new wing structure configuration or a new kind of connection. In fact, even market surveys of new products—where samples are mailed out or stuffed into sacks in the Sunday papers—can be conceived of as proof-of-concept tests that test the receptivity of a target market to a new product.

Proof-of-concept tests are scientific endeavors. We set out reasoned and supported hypotheses that are tested and then validated or disproved. Turning on a new artifact and seeing whether or not it "works" is not a proper proof-of-concept demonstration. An experiment must be designed with hypotheses to be disproved if certain outcomes result. Remember that prototypes and models differ in their underlying "reasons for being" and in their testing environments. While models are tested in controlled or laboratory environments, and prototypes are tested in uncontrolled or "real world" environments, the tests are *controlled* tests in both cases. Similarly, when we are doing proof-of-concept tests, we are doing controlled experiments in which the failure to disprove a concept may be key.

For example, suppose we had chosen Mylar containers as our new beverage product and we're designing them to withstand shipping and handling, both in the manufacturing plant and in the store. If we think of all of the things that could go awry (e.g., stacks of shipping pallets that could topple) and analyze the mechanics of what happens in such incidents, then we might conclude that a principal design criterion is that the Mylar containers should withstand a force of X Newtons. We would then set up an experiment in which we apply a force of X N, perhaps by dropping the containers from a properly calculated height. If the bags survived that drop, we could say that they'd likely survive shipping and handling. However, we could not absolutely guarantee survival because there is no way we could completely anticipate every conceivable thing that might happen to a beverage-filled Mylar container. On the other hand, if the Mylar container fails a properly-designed drop test, we can then be certain that it will not survive shipping and handling, and so our concept is disproved. The National Aeronautics and Space Administration (NASA) conducted a similar proof-of-concept test for gas-filled shock absorbers for the Mars lander. There are potential issues of legal liability involved in product testing—for example, for how much nonstandard use of a product can a manufacturer be held responsible? These issues are beyond our scope.

Prototypes, models, and proof-of-concept testing have different roles in engineering design because of their intents and test environments. These distinctions must be borne in mind while planning for the design process.

5.7 GENERATING AND EVALUATING IDEAS FOR THE XELA-AID DESIGN PROJECT

When we left our chicken coop teams in Chapter 4, they had listed the functions that a successful coop must perform and developed metrics to assess their designs against project objectives. (Some of the metrics were not appropriate to a one-semester project, so one team reduced its metrics to a set of component tests.) Now we will look at some of the tools used by the teams, highlighting their use and some instances where they could have been used more effectively.

The teams used morphological charts to translate the identified functions into meaningful alternatives. One team used a single, very long morphological chart, which here is broken into two parts in Figures 5.7 and 5.8. Primary functions are indicated in boldface, while alternative subfunctions are also listed in the functions column in normal type. For example, the primary function allow human entry/exit of coop includes means such as design that does not involve entering, side-hinged door(s), top hinged door(s), and removable roof that could be reached over. Alternative means for keeping the coop door closed, gathered under the means latch on door, itself a specific means of allowing entrance into and exit from the coop, include hasp, hook and eye, and board across doors. Note the difference in scale between these two lines of the chart. While a decision on how to keep a door latched must be made, it is certainly a different kind of decision than whether or not to have a door at all, or whether to have a removable roof. The difference depends on whether we are doing conceptual design or detailed design. Confusing and combining these two types of design has the unfortunate effect of cluttering up an important conceptual decision. A further problem to consider in the case of a very long chart such as that depicted in Figures 5.7 and 5.8 is that the number of possible combinations is very large. We saw in Section 5.3 that our beverage container design, which had only a few functions, had 864 possible outcomes. For this coop, the total number of outcomes is overwhelming. Clearly, a strategy is needed for grouping and organizing the functions and the resultant design alternatives.

The second team broke the functions into general areas, as shown in Figures 5.9 (a)–(d). In this case the team organized the functional decisions in terms of safety of inhabitants, egg production, growth of chicks, and maintenance of adult birds. This allows each morph chart to have only three or four functions, which results in more manageable alternatives. In some cases, the alternatives that are considered for one of the functional areas may also apply to others, which does constrain the number of alternatives that must ultimately be considered. For example, egg incubation and brooding of chicks are clearly

FUNCTION	POSSIBLE MEANS					
Allow for egg collection	have a sloped coop floor that sort of spirals down and eggs collect in a bin	leave where laid, allow women to pick up	ramp from each nest going to same place	nests bucket area underneath	conveyor ramp	
Protect eggs	hay in nests, if left in nests	nothing in nests, if eggs are left in nests	pads along "egg route" if not left in nests	calcium in diet (this works with others simultaneously)		
Allow incubation of eggs	villagers (in greenhouses)	villagers (in coop)	allow chickens			
Ventilate coop	open walls	gaps in top of solid walls, protected by overhang	"windows" in walls	fans powered by wind		
Nests	one communal	several communal	individual			
Allow for nest cleaning	wire bottom, so droppings fall through	removable nests that could be taken out for cleaning	non-removable, just go in and clean	line nest with hay, replace that		
Allow for chicken entry	doggie type door	use human entry door, keep closed during the day	use human entry door, keep open during the day	make small entry that doesn't close		
Allow for waste removal	wire-mesh floor that allows waste to fall through	coop that could be lifted off base, allowing for collection	dirt floor, swept	concrete floor, swept	wood floor, swept	potty train them
Allow human entry/exit of coop	design that does not involve entering	side-hinged door(s)	top-hinged door(s)	removable roof that could be reached over		
Allow human entry/exit of perimeter	gate in fence	no fence	climb over fence			

FIGURE 5.7 This is the first half of a morph chart developed by one of the student teams designing a chicken coop for a Guatemalan village cooperative. Note that it is quite extensive (as is Figure 5.8). This completeness is a virtue in analysis, but it is likely to result in too many combinations for effective design selection.

POSSIBLE MEANS

FUNCTION							
Protect chickens from predators during day	perimeter fence that does not go underground	perimeter fence that does go underground	underground coop				
Protect chickens from predators during night	no added protection for night	strong coop inside perimeter for night use only	underground coop				
Protect chickens from weather	sloped roof, no walls	sloped roof, on walls	flat roof, no walls	flat roof, on walls			
Latch on door	hasp	hook and eye	board across doors				
Walls	solid walls of wood or bricks	no walls chicken wire walls					
Size of perimeter fence	high	low					
Material for perimeter fence	chicken wire	wood					
Material for perimeter posts	wood stakes	rebar, double or triple bound	U-posts (or T)				
Roof	metal	corrugated fiberglass	corrugated metal	wood			
Floor (?)	concrete	dirt					
Allow for chicken feeding	wood trough, movable	wood trough, fixed to wall	"feeding room" so mess is localized	bottom of trash can or similar cylinder	individual bowls	concrete trough	container with sides tapered
Food place	carport like roof overhang	inside coop	outside coop				
Allow for chicken watering	wood trough, movable	wood trough, fixed to wall		bottom of trash can or similar cylinder	woven individual bowls	woven group bowls	concrete trough
Keep water fresh	slow running water	drain water through bottom by some sort of hole on coop floor	separate from food to prevent it from getting wet				

FIGURE 5.8 This is the second half of the morph chart started in Figure 5.7. In this chart, the team has offered alternatives to a number of the features that depend on previous choices, such as the size of the perimeter fence (if there is one), and the material for the roof. Once again, the team might have made better decisions had it separated the basic functions from the secondary functions. Which are basic functions and which are secondary?

(a) Safety of inhabitants

FUNCTION	MEANS			
Predator exclusion	Metal skirt on posts	Concrete floor, walls	Sand or gravel floor	Extend fence below ground
Waste removal	Waste pit	Dropping boards	Waste falls through sunfloor	Treat waste with lime
Climate control	Chicken wire walls	Burlap covered chicken wire	Windows	Sheet metal roof
Prevention of intra-coop violence	Separate chicks from adults	Separate turkeys, chickens	Round off interior corners	

(b) Egg Production

FUNCTION	MEANS			
Egg collection	Dark nest boxes	Communal nests	Cages	
Egg incubation	Broody hens	Box with heat lamp	In humans' homes	
Separation of infertile (eating) eggs	By hand during incubation (candling)			

(c) Growth of chicks

FUNCTION	MEANS			
Brood chicks	Broody hens	Box with heat lamp	In humans' homes	
Water chicks	Troughs	Fountains (inverted jar or tray)	Collect rain into troughs or basins	
Feed chicks	Communal troughs filled daily by hand			

(d) Maintenance of adult birds

FUNCTION	MEANS			
Promote growth	Allow access to insects	Mix ground meat into feed (byproducts)	Exercise area	Ground sea-shells in food
Water	Troughs	Fountains (inverted jars or trays)	Collect rain into troughs or basins	
Feed	Communal troughs filled daily by hand			

FIGURE 5.9   These are morph charts from another team that designed a chicken coop for the Guatemalan village cooperative. Note that this team has organized the functions into several categories to simplify its decision making. Notice also that many of the "functions" could be easily reworked into the standard "verb-object" syntax that we discussed in Chapter 4.

related in terms of the role of the broody hens because it is unlikely that both egg incubation and broody hens would be used to raise the young chicks.

The teams could have generated a large number of design alternatives from various combinations. However, many of these are just minor variants that are not conceptually different. In fact, the primary conceptual alternatives developed could be characterized in terms of three basic design sets: cages for containing the birds while they grow and mature,

The Outer Structure

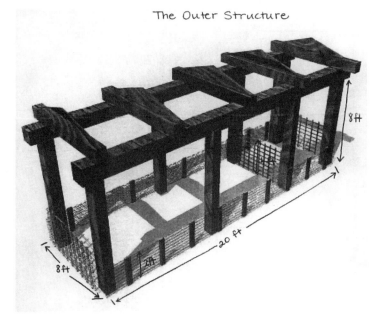

FIGURE 5.10 This is a sketch of the outer structure for a chicken coop design proposed by one of the the student teams. The team has chosen to enclose the entire space and then fill that area with cages and nesting areas, following the example of U.S. poultry factories. The size of the beams used to support the structure suggest that the outer shell can also be used as a railroad bridge if properly stabilized.

much like those used in high-production chicken farms in the U.S.; a fenced-in area that encloses a small building for the chicks and nests; and a fenced-in area containing separate areas for nests (egg production) and for roosting. Different versions of these three alternatives were explored by both teams, including greater and lesser dependence on villager labor, different roofing and fencing ideas, and different construction materials. Within their conceptual designs, each team also considered several more detailed design variations.

Both teams used simplified versions of the evaluation instruments proposed earlier in this chapter, using a verbal version of the best of class method to rank alternatives and weight them by the relevant objectives. This resulted in distinctly different results for the two teams.

One of the teams selected a coop design in which all of the available space would be used to build enclosing cages and a limited number of nestboxes. Figures 5.10 and 5.11 show sketches for this design choice. The cages are made largely of rebar and chicken wire with supporting posts of concrete block. They were designed and built to stack two levels high, which allowed the team to use all of the available space for production and would result in a capacity of 64 chickens and 12 or more chicks. However, the team estimated that the cost of this design would be quite high.

The other team selected a $6 \times 3 \times 4$ (ft) coop structure surrounded by a wire fence, shown in Figure 5.12, which was to be built of locally available concrete blocks. It would be divided into two halves: one for nests, one for roosting. The structure would be surrounded by a gated fence buried one foot underground to keep out predators. This design was intended to support 50 chickens and be cheaper than other team's designs, but it was not very cheap in absolute terms.

Why did two teams in the same class devise such different conceptual designs for the same client? The answer can be seen in Figures 3.9 and 3.10, wherein each team attached

FLOOR PLAN:FINAL DESIGN

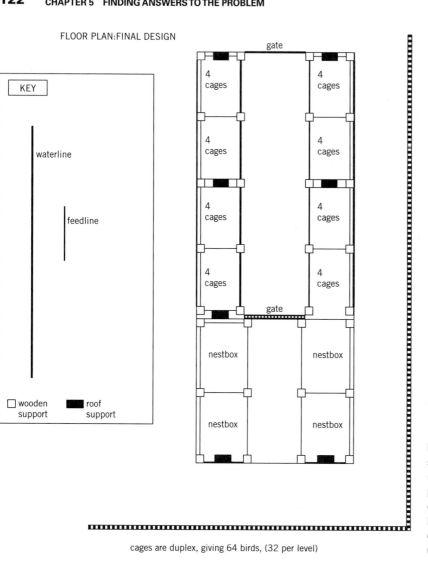

cages are duplex, giving 64 birds, (32 per level)

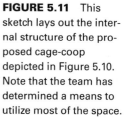

FIGURE 5.11 This sketch lays out the internal structure of the proposed cage-coop depicted in Figure 5.10. Note that the team has determined a means to utilize most of the space.

differing weights to the client's objectives. One team weighted increased production as more than one-half of the overall objectives (55%), while the team rated this at only 11%, focusing instead on matters such as the safety of inhabitants (28%), survival in the Guatemalan climate (25%), and ease of repair (14%). Thus, one team placed a very high premium on the number of chickens that could be supported, even at the expense of higher initial costs and greater design complexity, while the other team emphasized ease of repair or stability against natural forces. In Chapter 6 we will see which of the conceptual designs was ultimately adopted by the villagers and Xela-Aid.

It is instructive to report how the teams modeled and tested their design choices. Both teams faced a common problem that all designers must address: How can a design be confirmed without the time and budget resources to build and operate a functioning chicken coop? The cage team wanted to establish the key concept that safe, usable cages could be

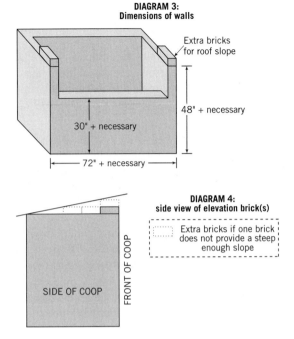

DIAGRAM 3:
Dimensions of walls

Extra bricks
for roof slope

48" + necessary

30" + necessary

72" + necessary

DIAGRAM 4:
side view of elevation brick(s)

Extra bricks if one brick
does not provide a steep
enough slope

FRONT OF COOP

SIDE OF COOP

FIGURE 5.12 Another student team's approach to the coop problem was to fence in the entire available space, and use a smaller building within that fence to provide shelter and nesting areas for the birds. This structure has a lower wall in the front that acts to support nestboxes and egg drawers. It is not obvious that this function could not be met by other means in ways that would allow for easier cleaning and ventilation.

constructed with available materials. Therefore, they first constructed a stack of two cages, using concrete blocks, rebar, and wire. It found, somewhat to its dismay, that cages constructed this way would not only be very heavy; they would also be unstable without lateral bracing, which would be quite difficult and complicated. (The team's basic hypothesis was that the cages would resist moderate lateral forces, defined in terms of one person pushing with moderate force on the structure. In fact, such forces caused a great deal of "sway" and thus served to *disprove* the team's hypothesis.) The team's final design thus substituted a wood frame as a support for the cages of rebar and chicken wire. By measuring the construction time of a complete cage of available materials, the team also learned that building the cages as originally designed would be very complicated and time consuming; simplifications were needed. Once again, the hypothesis that the cages could be built as designed within the time specified was disproved. These test results that failed to live up to expectations and led the team to modify and improve its design, show how testing can be used to identify both design shortcomings and possible fixes.

The second team also tested certain components of its design prior to completing and delivering it to the client. Here the ability to build a safe and strong fence using materials known to be locally available was a key element in the design. The team considered several alternatives, including using single pieces of rebar for the posts, doubled pieces, and triple pieces. It obviously is cheaper to use fewer pieces, and easier to construct. The team learned through experimentation that triple rebar, wired together at intervals of 18 in. or less, could support both the weight of the fence and the team's estimates of the lateral forces. So, once again, initially disappointing experiments led the team to a better design—which is the point of proof-of-concept testing.

5.8 MANAGING THE GENERATION AND SELECTION OF DESIGN ALTERNATIVES[1]

We have consistently emphasized the importance of team dynamics and, especially, respect for the ideas of others. We have done this because the interpersonal aspects of how a team uses the formal tools and techniques are as important as the tools and techniques themselves. However, it is also important to consider the roles of some of the more formal management tools provided in Chapter 7.

First, the amount of space devoted to describing the management activities should not be taken as an accurate estimate of the actual time that a team will need to conduct them. Idea generation activities often proceed relatively quickly, while demonstrating proofs of concept will take much more time. This is a natural consequence of a team's investment in thought and study about the problem. While formal alternative generation may begin at a specific time, most of the team members will have thought about alternatives throughout the project. Evaluation of the alternatives, on the other hand, requires that each alternative be laid out systematically, and so evaluation cannot be distributed over the life of the project. Proof-of-concept testing of an idea is likely to be a "time sink" for many design teams since it often requires team members to apply skills they might not know they have, and to order parts and tools that may take time to arrive. Thus, concept testing may require several iterations to "make it work," so a team should revisit its schedule and plans, both to update them and to determine whether its goals can actually be realized.

Second, the activities described below offer real opportunities for design teams to consider how responsibility for these activities is allocated among team members. Brainstorming and similar creative activities are most effective when the entire team is present, while a physical model may be best built by a subset of the team. Hard feelings can result if such staffing issues are not handled properly.

Finally, design teams almost always work under the pressures of fixed deadlines and other competing activities. This places a premium on time management: Adjustments should be made early, as soon as information permits. There is no substitute for having a team review its work plan at regular intervals and change it as needed.

5.8.1 Task management

At the start of a design project, a team may be sufficiently unfamiliar with the particular design problem being solved, and perhaps design processes in general, that it may have given only *pro forma* consideration to the tasks its members must perform to do a successful design. As we will detail in Chapter 7, a team should build a work breakdown structure (WBS) to lay out these tasks and the times each is anticipated to take. The team may not have thought through the implications and consequences of the time assignments made. However, by the time a team has generated some alternative concepts, it should have a much better understanding of its problem and the tasks it must undertake to complete the project. For this reason, the team should, at a minimum, review its initial WBS as it generates—before it evaluates—design alternatives, well before proceeding with proof-of-concept testing and prototyping. Testing and prototyping will be useful to a team *only after*

---

[1] This section anticipates more detailed descriptions of management tools presented in Chapter 7.

its design concepts have been properly evaluated. Among the tasks the team needs to consider are:

- selecting functions and actions of the prototype or physical model,
- acquiring parts or special materials needed to build the prototype,
- building the prototype,
- developing a test protocol,
- conducting the test, and
- revising and reworking when the prototype doesn't work initially—and it almost never does!

Teams often find out at this late time in the project that their prototype or proof of concept will not be as complete as originally hoped. This ugly fact must be faced and considered in terms that allow a client or a later designer to build on the work already done. The results may be useless without this perspective.

5.8.2 Scheduling

Just as the tasks of the project require review, and perhaps revision, the schedule under which they can be accomplished should also be reviewed during the design generation, alternative evaluation, and testing stages. Depending on the time remaining to complete the project and the complexity of the tasks yet to be done, one of the methods proposed in Chapter 7 should be applied.

Teams that have been using a computer-generated *activity network* or some other standard project tool can easily modify their remaining activities and schedules. Two noteworthy points are that the team's responsibilities under this project must be reviewed, as must any other specific commitments that members may have that may affect the project. Team members need to think through how their personal commitments—both on the job and off, both schoolwork and extracurricular—might affect their ability to devote time to the project. Most software packages permit users to specify availability calendars for each team member. Such calendars should be updated as new information (e.g., new projects) becomes available. Most teams that do recalculate schedules on the computer are both surprised and dismayed. However, even if a revised schedule is discouraging, it is not an excuse for disregarding the schedule. Rather, it should be used as a prompt for discipline in adhering to the schedule.

For want of a selection the order was lost. . .

For want of an order the part was lost. . .

For want of a part the test was lost. . .

Teams that used simpler calendars to plan activities would now pay a price for their previous convenience. Clearly, a simple calendar cannot calculate anything, so a team must revisit its commitments and their implications. This may cause the team to set target dates that require a number of "all-nighters" or other unplanned or undesirable activities. In that case, the team has to "sweat the details" of scheduling, probably at every team meeting.

With either approach, a team must consider both logical and temporal relationships between activities. A prototype can't be assembled if its parts haven't arrived. Parts don't arrive if they haven't been ordered. Parts can't be ordered if they haven't been selected. Ignoring obvious orderings of events has been the pitfall of many design teams, in professional firms and in academic projects. Such errors can be avoided with planning and "a little clear thinking."

5.8.3 Budgeting

Closely related to the above discussion is the need for a team to create and use an appropriate budget. Many projects have limited financial resources. This must be considered before a team develops all of its metrics or its methods of proving a concept. It is also important to keep in mind that the budget for building a prototype or conducting a proof-of-concept experiment should not be confused with the budget and costs of building the final product. That distinction will be explored in Chapter 7, but it is important to realize that design teams are almost always faced with resource limits in establishing their design choices. Thus, teams will not be able to prove all of their key ideas within their available budgets or resources. In such cases, teams must set priorities and decide which concepts must be demonstrated by building prototypes or models, and which concepts can be established by analysis or reference to appropriate authority.

Once a team has agreed on how to spend its available resources it should devise a simple process for approving and making expenditures. Sometimes a team's parent organization will have formal procedures (e.g., purchase orders, reimbursement forms, etc.), but even here someone on the team should have the responsibility for tracking and recording all expenditures. This helps avoid later problems with budget overruns or disapproved expenditures.

5.8.4 Monitoring and controlling progress

All of the foregoing discussion in this section has presumed that the team has a way to monitor and track its progress. We will discuss tools for this Chapter 7, but it is worthwhile to briefly describe how such tools can be translated into team-level success.

The single most important element in monitoring and tracking a team's activities relative to its plan is explicit, formal communication about these topics at team meetings. Every team meeting should include a brief review of the recent and near-future elements of the work plan, a discussion of progress to date, and consideration of any changes to the time allowed for activities.

Team reviews of recent, current, and future activities allow each member of the team to see what is being done, report on his or her work to date, and to identify her responsibilities and commitments. It is clearly a good thing to disseminate information this way within the team. It also tends to encourage members who may have been doing less work to do more, if only because they want to avoid saying: "Nothing to report." Such a review of activities should be a scheduled part of the team meeting agenda, either at the outset as a means of bringing everyone up to speed on team progress, or as the final agenda item, as a method of stimulating further work.

Discussion of progress to date should, wherever possible, focus on concrete or quantifiable goals and targets. It should avoid general remarks like, "We're coming along on that." Ideally, tasks should be examined in terms of specific deliverables or milestones. Alternatively, it may be useful to work toward a specific percentage of work completed on a task. It may seem artificial initially to force a responsible team member to indicate that a task is 25, 50, or 75 percent complete, but it can quickly become a group norm that allows members to gauge their own progress and to request help when it's needed.

After discussing the progress made to date and the tasks yet to be done, a team should examine the implications of its current situation in the light of those tasks still undone. This should be done at or near the end of a meeting, since new tasks or modifications may have emerged during the meeting. Team leaders should allocate some time at the end of each meeting for this, preferably as part of a final agenda item that includes progress assessment. It is a mistake to try to update schedules and timelines as everyone is picking up their coat and heading for the door! An important element of this final updating is that someone on the team take responsibility for sending out (perhaps by e-mail) a revised schedule and *percent-complete matrix* (PCM) that reflects the project status as of this meeting. This allows members to review their commitments and affords the team a record of its progress as the project unfolds.

Poorly managed teams do even more work.

  At first, all of these management tasks may seem to "get in the way" of the engineering and the design. While managing the project may seem like a lot of work, experience shows that unmanaged teams do even more work. In fact, unmanaged teams usually do things more than once. There is an old saying to the effect that teams rarely accomplish more than they plan on. This is certainly true of most design teams.

5.9 NOTES

Section 5.2: The address of the USPTO's Web site is www.uspto.gov. Another often-used Web site is www.ibm.com/patents. Group methods of idea generation are explored and described in (Shah, 1998) and synectics are defined in Webster's Ninth New Collegiate Dictionary (Mish, 1983) and described in (Cross, 1994). Approaches to creativity and analogical thinking in a group setting are described in (Hays, 1992).

Section 5.4: Zwicki (1948) originated the idea of a morphological chart. Further discussion and examples of morph charts can be found in (Cross, 1994), (Jones, 1992), and (Hubke, 1988).

Section 5.5: Pugh's concept selection method is discussed in (Pugh, 1990), (Ullman, 1992, 1997), and (Ulrich and Eppinger, 1995, 2000).

Section 5.7: The results from the Xela-Aid chicken coop design project are taken from final reports (Gutierrez, 1997) and (Connor, 1997) submitted during the Spring 1997 offering of Harvey Mudd College's freshman design course, E4: Introduction to Engineering Design.

5.10 EXERCISES

5.1 Explain what is meant by the term "design space" and discuss how the size of the design space might affect a designer's approach to an engineering design problem.

5.2 Using the functions developed in Exercise 4.4, develop a morphological chart for the portable electric guitar.

5.3 Organize and apply a process for selecting means for realizing the design of the portable electric guitar.

5.4 Using Web-based patent lists (identified in Section 5.8), develop a list of patents that are applicable to the portable electric guitar.

5.5 Using the functions developed in Exercise 4.6, develop a morphological chart for the rainforest project.

5.6 Organize and apply a process for selecting means for realizing the design of the rainforest project.

5.7 Using Web-based patent lists (identified in Section 5.8), develop a list of patents that are applicable to the rain forest project.

5.8 Describe an acceptable proof of concept for the rainforest project. Would a prototype be appropriate for this project? If so, what would be the nature of such a prototype?

CHAPTER 6

REPORTING THE OUTCOME

How do we let our client know about the solutions?

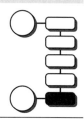

REPORTING IS an essential part of a design project: The project is not complete if its results have not been communicated to the client and other stakeholders designated by the client. Final design results can be communicated in several ways, including oral presentations, final reports (that may include design drawings and/or fabrication specifications), and prototypes and models such as those discussed in Section 5.6. In this chapter we first consider some common guidelines for all reporting modes, and then we look at final technical reports, oral presentations, and design drawings.

Independent of the details, however, note that the primary purpose of such communication is to inform the client *about the design*, including explanations of why this design was chosen over competing design alternatives, and how. It is most important to convey the *results* of the design process. The client is likely uninterested in the history of the project or in the design team's internal workings. Thus, final reports and presentations are *not* chronologies of a team's work. Rather, they should be lucid descriptions of design *outcomes*.

6.1 GENERAL GUIDELINES FOR TECHNICAL COMMUNICATION

There are some basic elements of effective communication that apply to writing reports, giving oral presentations, and providing informal updates to your client. Thomas Pearsall summarized these common concepts as the seven principles of technical writing (see Figure 6.1), but they clearly apply more generally. And while Pearsall devoted more than one-half of his book to these principles, we will summarize them here as a prelude to the rest of this chapter.

Know your purpose.

This is the writing analogue of understanding objectives and functions for a designed artifact. Just as we want to understand what the designed object must be and must do, we need to understand the goals of a report or presentation. In many cases design documentation seeks to inform the client about the features and design elements of a selected design. In other cases the design team may be trying to persuade a client that a design is the best alternative. In still other cases a designer may wish to report how a design operates to users,

> 1. Know your purpose.
> 2. Know your audience.
> 3. Choose and organize the content around your purpose and your audience.
> 4. Write precisely and clearly.
> 5. Design your pages well.
> 6. Think visually.
> 7. Write ethically!

FIGURE 6.1 Pearsall's seven principles of effective technical writing. These same principles apply to all modes of communication, including reports, memos and presentations. (Adapted from Pearsall, 2001.)

whether beginners or highly experienced. You may not realize any purpose if you don't know what purpose your writing or presentation is serving.

Know your audience.

We have all sat through lectures where we didn't know what was going on or where the material was so simple that we already knew it. We can often take some action once we realize that the material is not set at a level that we find appropriate. Similarly, when documenting a design, it is essential that a design team structure its materials to its targeted audience. Thus, the team should ask questions like, "What is the technical level of the target audience?" and "What is their interest in the design being presented?" Taking time to understand the target audience will help ensure that its members appreciate your documentation. Sometimes you may prepare multiple documents and briefings on the same project for different audiences. For example, it is quite common for designers to close projects out with both a technical briefing and a management briefing. It is also common for designers to confine calculations or concepts that are of limited interest to a report's primary audience to specific sections of their reports, usually appendices.

Choose and organize the content around your purpose and your audience.

Once we are sure of the purpose of the report or presentation and its target audience, it only makes sense to try to select and organize its content so that it will reach its intended target. The key element is to structure the presentation to best reach the audience. In some cases, for example, it is useful to present the entire process by which the design team selected an alternative. Other audiences may only be interested in the outcome.

There are many different ways to organize information, including: going from general overviews or concepts to specific details (analogous to deduction in logic); going from specific details to general concepts (analogous to induction or inference); sequencing design events chronologically (which we do not recommend); and describing devices or systems.

Once an organizational pattern is chosen, no matter which form is used, the design team should translate it into a written outline. As we will discuss below, this allows the team to develop a unified, coherent document or presentation and avoid needless repetition.

Write precisely and clearly.

This particular guideline sounds like "use common sense," that is, do something that everyone wants to do but few achieve. There are, however, some specific elements that seem to occur in all good writing and presentations. These include effective use of: short paragraphs (and other structural elements) that have a single common thesis or topic; short, direct sentences that contain a subject and a verb; and active voice and action verbs that allow a reader to understand directly what is being said or done. Opinions or viewpoints should be clearly identified as such. These elements of style should be learned and applied.

Young designers may have practiced these skills more in humanities and social science classes than in technical courses. This is acceptable, and even welcome, so long as the designer remembers that the goals of both technical and non-technical communications remain the same.

Design your pages well.

Whether writing a technical report or organizing supporting materials for a verbal briefing or presentation, effective designers utilize their media wisely. In technical reports, for example, writers judiciously use headings to support and extend the organizational structure of the report. A long section divided into several subsections helps readers understand where the long section is going, and it sustains their interest over the journey. Selecting fonts to highlight key elements or to indicate different types of information (such as new, important terms) guides the reader's eye to key elements on the page. White space on a page helps keep readers alert and avoids a forbidding look in documents.

Similarly, careful planning of presentation support materials such as slides and transparencies can enhance and reinforce important concepts or elements of design choices. Fonts that are large enough for the entire audience to see is an obvious, but often overlooked aspect of presentations. Just as white space on the page invites readers to enter the text, simple and direct slides encourage readers to listen to the speaker without being distracted visually. Presenters often fill slides with so many words or other content that their audience is forced to choose between the slide and speaker; this scenario should be avoided.

Think visually.

By their very nature, design projects invite visual thinking. Designs often start as sketches, analyses often begin with free body or circuit diagrams, and plans for realizing a design involve graphics such as objectives trees and work breakdown structures. Just as designers often find that visual approaches are helpful to them, audiences are helped by judicious use of visual representation of information. These can range from the design tools discussed throughout this book, to detailed drawings or assembly drawings to flow charts and cartoons. Even tables present an opportunity for a design team to concentrate attention on critical facts or data. In fact, given the enormous capabilities of word processing and presentation graphics software, there is no excuse for a team not to use visual aids in its reports and presentations. On the other hand, a team should not be seduced by their graphics' capabilities by, for example, clouding their slides with artistic backgrounds that make the words illegible. The key to success here, as it is with words, is to know your purpose and your audience, and to use your medium appropriately.

Write ethically!

Designers often invest themselves in the design choices they make, in time, effort, and even values. It is, therefore, not surprising that there are temptations to present designs or other technical results in ways that not only show what is favorable, but that suppress unfavorable data or issues. Ethical designers resist this temptation and present facts fully and accurately. This means that that *all* results or test outcomes, even those that are not favorable, are presented and discussed. Ethical presentations also describe honestly and directly any limitations of a design. Further, it is also important to give full credit to others, such as authors or previous researchers, where it is due. (Remember that this discussion of the seven principles began with an acknowledgement to their originator, Thomas Pearsall, and that each chapter of the book ends with references and citations.) We will talk more about engineering ethics in Chapter 9, where will also describe a landmark case involving an engineer widely acclaimed for both his work and for his ethics.

Now we turn our attention to specific forms of documentation.

6.2 THE PROJECT REPORT: WRITING FOR THE CLIENT, NOT FOR HISTORY

The usual purpose of a final or project report is to communicate with the client in terms that ensure the client's thoughtful acceptance of a team's design choices. The client's interests demand a clear presentation of the design problem, including analyses of the needs to be met, the alternatives considered, the bases on which decisions were made, and, of course, the decisions that were taken. The results should be summarized in clear, understandable language. Highly detailed or technical materials are often placed in appendices at the end of the report, in order to support clarity. In fact, it is not unusual, and in large public works projects it is the norm, for all of the technical and other supporting materials to be moved to separate volumes. This is especially important when the client and the principal stakeholders are not engineers or technical managers, but perhaps members of the general public.

The process of writing a final report, like so much of design, is best managed and controlled with a structured approach. The design process and report writing are strikingly similar, especially in their early, conceptual stages. It is very important to clearly delineate objectives, both for the designed object and for the project report. It is very important to understand the "market," that is, to understand both the user needs for the design and the intended audience of the final report. It is very important to be reflective and analytical and to recognize that analysis isn't limited to applying known formulas. We have described several tools that can support our thinking during the design process. Similarly, writing is also an analytical thinking process.

As with the design process, structure is not intended to displace initiative or creativity. Rather, we find that structure can be help us learn how to construct an organized report of the design results. In this case, one structured process that a design team might follow would include the following steps:

- determine the purpose and audience of the technical report;
- construct a rough outline of the overall structure of the report;
- review that outline within the team and with the team's managers or, in case of an academic project, with the faculty advisor;
- construct a topic sentence outline and review it within the team;
- distribute individual writing assignments and assemble, write and edit an initial draft;
- solicit reviews of the initial draft from managers and advisors;
- revise and rewrite the initial draft in response to its reviews; and
- prepare the final version of the report and present it to the client.

We now discuss the remaining steps in greater detail.

6.2.1 The purpose of and the audience for the final report

We have already discussed determining the purpose and audience of the report in general terms. Several points should be noted in the case of a final report. The first is that the report is likely to be read by a much wider audience than simply by the client's liaison with whom the team has been interacting. In this respect, the team needs to determine whether or not the liaison's interests and levels of technical knowledge are representative of the audience for the final report. The liaison may, however, be able to guide the team to a better understanding of the expected reader(s), and may highlight issues that may be of particular concern.

Another important element here is for the team to understand what the report's recipient hopes to do with the information in the final report. If, for example, the intent of the project was to create a large number of conceptual design alternatives, the audience is likely to want to see a full presentation of the design space that was explored. If, on the other hand, the client simply wanted a solution to a particular problem, he is much more likely to want to see how well the selected alternative meets the specified need.

A project report often has several different audiences, in which case the team will have to organize information to satisfy each of these target groups. This may include using technical supplements or appendices, or it may call for a structure that begins with general language and concepts, and then explores these concepts in technical subsections. The team, however, should write clearly and well for each audience, regardless of the organizational principle selected.

6.2.2 The rough outline: Structuring the final report

Only a fool would start building a house or an office building without first analyzing the structure being built and organizing the construction process. Yet many people sit down to prepare a technical report and immediately begin writing, without trying to lay out in advance all of the ideas and issues that need to addressed, and without considering how these ideas and issues relate to each other. One result of such unplanned report writing is that the report turns into a project history, or worse, it sounds like a "What I Did Last Summer" essay: First we talked to the client, then we went to the library, then we did research, then we did tests, etc., etc., etc. While technical reports may not be as complex as high-rise office buildings or airplanes, they are still complicated enough that they cannot simply be written as simple chronologies. Reports must be planned!

The first step in writing a good project or final report is building a good rough outline for the report's overall structure. That is, we identify the major sections into which the report is divided. Typically, some of these sections are:

- Abstract
- Executive summary
- Introduction and overview
- Analysis of the problem, including relevant prior work or research
- Design alternatives considered
- Evaluation of design alternatives and basis for design selection
- Results of the alternatives analysis and design selection
- Supporting materials, often set out in appendices, including:
  - Drawings and details
  - Fabrication specifications
  - Supporting calculations or modeling results
  - Other materials that the client may require

This outline looks like a table of contents, as it should, because a final report of an engineering or design project must be organized so that a reader can go to any particular section and see it as clear and coherent stand-alone document. It is not that we think things should be taken out of context. Rather, it is that we expect each major section of a report to make sense all by itself; that is, it should tell a complete story about some aspect of the design project and its results.

Having identified a rough outline as the starting point for a final report, *when* should that outline be prepared? Indeed, when should the final report be written? It evident that we can't write a *final* report until we have completed our work and identified and articulated a final design. On the other hand, as with the design process, it is very helpful to have an idea of where we're going with a final report so that we can organize and assemble it along the way. It can be very helpful to develop a general structure for the final report early in project. We can then track and appropriately file or label key documents from the project (e.g., research memoranda, drawings, objectives trees) according to whether their contents would appear in the final report. Thinking about the report early on also emphasizes thinking about a project's *deliverables*, that is, those items that the team is contracted to deliver to the client during the project. Organizing the final report early on may make the final stages or endgame of the project much less stressful, simply because there will be fewer last-minute things to identify, create, edit, and so on, so they can be inserted into the final report.

6.2.3 The topic sentence outline: Every entry represents a paragraph

A cardinal rule of writing states that *every single paragraph* of a piece should have a topic sentence that indicates that paragraph's intent or thesis. Once the rough outline of a report has been established, it is usually quite useful to build a corresponding, detailed *topic sentence outline* (TSO) that identifies the themes or topics that, collectively, tell the story told within each section of the report. Thus, if a topic is identified by an entry in the TSO, we can assume that there is a paragraph in which that topic is covered.

Each entry in a topic sentence outline corresponds to a single paragraph.

The TSO enables us to follow the logic of the argument or story and assess the completeness of each section being drafted, as well as of the report as a whole. Suppose there is only one entry in a TSO for something that we consider important, say, the evaluation of alternatives. One implication of this is that the final report will have only one paragraph devoted to this topic. Since the evaluation of alternatives is a central issue in design, it is quite likely that there should be entries on a number of aspects, including the evaluation metrics and methods, the results of the evaluation, key insights learned from the evaluation, the interpretation of numerical results—especially for closely rated alternatives, and the outcome of the process. Thus, a quick examination of this TSO shows us that a proposed report is not going to address all of the issues that it should.

For the same reason, of course, TSOs help identify appropriate cross-references that should be made between subsections and sections as different aspects of the same idea or issue are addressed in different contexts. The format of a TSO also makes it easier to eliminate needless duplication because it is much easier to spot repeated topics or ideas. In Section 6.5 we show examples of a TSO that demonstrates some of these points.

Writing in this way is hard, but TSOs do provide a number of advantages to a design team. One is that a TSO forces the team to agree on the topics to be covered in each section.

It quickly becomes clear if a section is too short for the material, or if one of the co-authors (or team members) is "poaching" on another section that was agreed upon in the rough outline. Another advantage of a good TSO is that it becomes easier for team members to take over for one another if something comes up to prevent a "designated writer" from actually writing. For example, a team member may suddenly find that the prototype is not working as planned and she needs to do some more work on it. TSOs also make life easier for the team's report editor (see the next section) to begin to develop and use a single voice.

Notwithstanding the definition of the abbreviation TSO, the entries in a TSO do not really have to be grammatically complete sentences. However, they should be complete enough that their content is clear and unambiguous.

6.2.4   The first draft: Turning several voices into one

The larger the writing team, the greater the need for a single editor.

One advantage of agreed-upon rough and topic sentence outlines is that their structure allows teams members to write in parallel or simultaneously. However, this advantage comes at a price, most notably that of corralling the efforts of several writers into a single, clear, coherent document. Simply put, the more writers, the greater the need for a single, authoritative editor. Thus, one member of the team should enjoy the rights, privileges and *responsibilities* pertaining to being the editor. Further, the team should designate an editor as soon as the planning of the report begins, hopefully at or near the onset of the project.

The editor's role is to ensure that the report flows continuously, is consistent and accurate, and speaks in a single voice. *Continuity* means that topics and sections follow a logical sequence that reflects the structure of the ideas in the rough outline and the TSO. *Consistency* means that the report uses common terminology, abbreviations and acronyms, notation, units, similar reasoning styles, and so on, throughout the report and all of its appendices. It also means, for example, that the team's objectives tree, pairwise comparison chart, and evaluation matrix all have same elements; if not, discrepancies should be noted explicitly and explained.

Accuracy requires that calculations, experiments, measurements, or other technical work are done and reported to appropriate professional standards and current best practices. Such standards and practices are often specified in contracts between a design team and its client(s). They typically provide that stated results and conclusions must be supported by the team's prior work. Accuracy, as well as intellectual honesty, also require that technical reports do not make unsupported claims. There is often a temptation in a project's final moments to add to a final report something that wasn't really done well or completely. This temptation should be avoided.

Good technical reports speak in a single voice.

The *voice* or style of a report reflects the way in which a report "speaks" to the reader, in ways very similar to how people literally speak to each other. It is essential that a technical report *speaks with a single voice*—and ensuring that single voice is one of the editor's most important duties. This mandate has several facets, the first of which is that the report has to read (or "sound") as if it was written by one person, even when its sections were written by members of a very large team. A president of the United States sounds like the same, familiar person, even while using several speechwriters. So, similarly, a technical report must read in a single voice. Further, that voice should normally be more formal and impersonal than the voice of this book. Technical reports are not personal documents, so they should not sound familiar or idiosyncratic. Also, that single voice can either be active or passive as modern practice renders both acceptable. It is important only that the voice of the report be the same from the opening abstract through the closing conclusions and to the last appendix.

Clearly, there are serious issues for the team dynamics of the writing process. Team members have to be comfortable surrendering control of pieces they have written, and they have to be willing to let the editor do her job. We will discuss aspects of the team dynamics of report writing in Section 6.6.1.

6.2.5 The final, final report: Ready for prime time

A good review process ensures that a draft final report gets thoughtful reconsideration and meaningful revision. Draft reports benefit from careful readings and reviews by team members, managers, client representatives or liaisons, faculty advisors, as well as people who have no connection with the project. This means that as we are trying to wrap up our project report, we need to incorporate reviewers' suggestions into a final, high-quality document. There are a few more points to keep in mind.

A final report should be professionally-done and *polished*. This does not mean that it needs glossy covers, fancy type and graphics, and an expensive binding. Instead, it means that the report is clearly organized, easy to read and understand, and that its graphics or figures are also clear and easily interpreted. The report should also be of reproducible quality because it is quite likely to be photocopied and distributed within the client's organization, as well as to other individuals, groups, or agencies.

We should also keep in mind that a report may go to a very diverse audience, not simply to peers. Thus, while the editor needs to ensure that the report speaks with a single voice to an anticipated audience, she should try as much as possible to also ensure that the report can be read and understood by readers that may have different skill levels or backgrounds than either the design team or the client. An *executive summary* is one way to address readers who may not have the time or interest to read all of the details of the entire project.

Finally, the final report will be read and used by client(s), who will, one hopes, adopt the team's design. This means that the report, including appendices and supporting materials, are sufficiently detailed and complete to stand alone as the final documentation of the work done. We say more about this in Section 6.4.

6.3 ORAL PRESENTATIONS: TELLING A CROWD WHAT'S BEEN DONE

Most design projects call for a number of meetings with and presentations to clients, users, and technical reviewers. Such presentations may be made before the award of a contract to do the design work, perhaps focusing on the team's ability to understand and do the job in the hope of winning the contract in a competitive procurement. During the project, the team may be called upon to present their understanding of the project (e.g., the client's needs, the artifact's functions, etc.), the alternatives under consideration and the team's plan for selecting one, or simply their progress toward completing the project. After a design alternative has been selected by the team, the team is often asked to undertake a design review before a technical audience to assess the design, identify possible problems, and suggest alternate solutions or approaches. At the end of a project, design teams usually report on the overall project to the client and to other stakeholders and interested parties.

Because of the variety of presentations and briefings that a team may be called upon to make, it is impossible to examine each of them in detail. However, there are elements common and key to most of them. Foremost among these are the needs to: identify the audience; outline the presentation; develop appropriate supporting materials; and practice the presentation.

6.3.1   Knowing the audience: Who's listening?

Design briefings and presentations are given to many types of audiences. For example, some projects call for the work to be reviewed periodically by technical experts. Others are concerned with a design's implications for management. Some may be concerned with how a design will be manufactured. Consider the new beverage container whose design we began in Chapter 3. Our design work might have to be presented to logistics managers who are concerned with how the containers will be shipped to warehouses around the country. The marketing department, concerned with establishing brand identity with the design, might want to hear about our design alternatives. Similarly, the manufacturing managers will want to be briefed about any special production needs. Thus, as noted in our review of Pearsall's seven principles, a team planning a briefing should consider factors such as varying levels of interest, understanding, and technical skill, as well as the amounts of available time. We can assume that most attendees at a meeting are interested in at least some aspect of a project, but it is generally true that most are only interested in particular dimensions of that project. A team usually can identify such interests and other dimensions simply by asking the organizer of the meeting.

Once the audience has been identified, a team can tailor its presentation to that audience. As with other deliverables, the presentation must be properly organized and structured: The first step is to articulate a rough outline; the second is to formulate a detailed outline; and the third is to prepare the proper supporting materials, such as visual aids or physical models.

6.3.2   The presentation outline

Just as with the final report discussed in Section 6.2, a presentation must have a clear structure. We achieve this structure by again developing a rough outline. This presentation structure and organization, which should be logical and understandable, then guides the preparation of supporting dialogue and discussion. And because a design presentation is neither a movie nor a novel, it should not have a "surprise ending." A sample presentation outline would include the following elements:

- A *title slide* that identifies the client(s), the project, and the design team or organization responsible for the work being presented.

- An *overview of the presentation* that shows the audience the direction that the presentation will take.

- A *problem statement*, including the initial statement given by the client and an indication of how that problem statement changed as the team came to understand the project.

- *Background material on problem*, including relevant prior work and other materials developed through team research.
- *The key objectives of the client and users* as reflected in the top level or two of the objectives tree.
- *Functions that the design must fulfill*, focusing on basic functions, but possibly including issues of unwanted secondary functions.
- *Design alternatives*, particularly those that were still considered at the evaluation stage.
- *Highlights of the evaluation procedure and outcomes*, including key metrics or objectives that bear heavily on the outcome.
- *The selected design*, explaining why this design was chosen.
- *Features of the design*, highlighting aspects that make it superior to other alternatives and any novel or unique features.
- *Proof of concept testing*, especially for an audience of technical professionals for whom this is likely to be of great interest.
- *A demonstration of the prototype*, assuming that a prototype was developed and that it can be shown. Videotapes or still photos may also be appropriate here.
- *The conclusion(s)*, including the identification of any future work that remains to be done.

There may not always be enough time to include all of these elements in a talk or presentation, so a team may need to limit or exclude some of them. This decision will also depend, at least in part, on the nature of the audience.

Once the rough outline has been articulated, a detailed outline (analogous to a TSO) of the presentation should also be developed. This is important both to ensure that the team understands the point that is being made at all times in the presentation and to develop corresponding bullets or similar entries in its slides or transparencies. Bullets generally correspond to entries in the detailed outline.

Preparing a detailed outline for the presentation may seem like a great deal of work, just as developing a TSO for a report appears at first to be cumbersome. And, ironically, team members with public speaking experience may be most resistant to such tasks, most likely because they have already internalized a similar method of preparation. However, since presentations represent the entire team, every member of a team should review the structure and details of their presentations, as well as the detailed outline required by such reviews.

6.3.3 Presentations are visual events

Just as the team needs to know the audience, it should also try to know the setting in which they will be presenting. Some rooms will support certain types of visual aids, while others will not. At the earliest stages of the presentation planning, the design team should find out what devices (e.g., 35 mm slide projectors, overhead projectors, computer connections) are available and the general setting of the room in which it will be presenting. This includes its size, capacity, lighting, seating, and other factors. Even if a particular device or setup is said to be available, it is always wise to bring along a backup, such as transparencies or foils to back up a slide presentation.

There are other tips and pointers to keep in mind about visual aids, including:

- Avoid using too many slides or graphics. A reasonable estimate of the rate of slides at which slides can be covered is 1–2 slides per minute. If too many slides are "planned," the presenter(s) will find themselves rushing through the slides in the hope of finishing. This makes for a far worse talk than a smaller selection wisely used.

- Beware of "clutter." Slides should be used to highlight key points; they are not a direct substitute for the reasoning of the final report. The speaker should be able to expand upon the points in the slides.

- Make points clearly, directly, and simply. Slides that are too flashy or clever tend to detract from a presentation.

- Use color skillfully. Current computer-based packages support many colors and fonts, but their defaults are often quite appropriate. Also, avoid weird or clashing colors in professional presentations.

- Do not reproduce completed design tools (e.g., objectives tree, large morph charts) to describe the outcomes of the design process. Instead, highlight selected points of the outcomes. This is a situation where it makes more sense to refer the audience to the final report for more detailed information.

Effective visuals don't replace effective speakers; they enhance them.

It is worth remembering that audiences tend to read visual aids as the speaker is talking. Therefore, the speaker does not need to read or quote those slides. The visual aids can be simpler (and more elegant) in their content because the visual aids reinforce the speaker, rather than the other way around.

Finally, if design drawings are being reproduced and shown, the size and distance of the audience must be considered carefully. Many line drawings are difficult to display and to see and interpret in large presentations.

6.3.4 Practice makes perfect, maybe . . .

Presenters and speechmakers are usually effective because they have extensive experience. They have given many speeches and made many presentations, as a result of which they have identified styles and approaches that work well for them. Design teams cannot conjure up or create such real-world experience, but they can practice a presentation often enough to gain some of the confidence that experience breeds. To be effective, speakers typically need to practice their parts in a presentation alone, then in front of others, including before a "rump" audience including at least some people who are not familiar with the topic.

Another important element of effective presentation is that speakers use words and phrases that are natural to them. Each of us normally has an everyday manner of speech with which we are comfortable. While developing a speaking style, however, we have to keep in mind that ultimately we want to speak *to* an audience in *their* language, and that we want to maintain a professional tone. Thus, when practicing alone, it is useful for a presenter to try saying the key points in several different ways as a means of identifying and adopting new speech patterns. Then, as we find some new styles that work, we should repeat them often enough to feel some ownership.

Practice sessions, whether solitary or with others, should be timed and done under conditions that come as close as possible to the actual environment. Inexperienced speakers

Speech coaches and athletic coaches say that you play how you practice!

typically have unrealistic views of how long their talk will last, and they also have trouble setting the right pace, going either too fast or too slow. Thus, timing the presentation—even setting a clock in front of the presenter—can be very helpful. If slides (or transparencies or a computer) are to be used in the actual presentation, then slides (or transparencies or a computer) should be used in practice.

The team should decide in advance how to handle questions that may arise. This should be discussed with the client or the sponsor of the presentation before the team has finished practicing. There are several options available for handling questions that are asked during a talk, including deferring them to the end of the talk, answering them as they arise, or limiting questions during the presentation to clarifications of facts while deferring others until later. The nature of the presentation and the audience will determine which of these is most appropriate, but the audience should be told of that choice at the start of the presentation. When responding to questions, it is often useful for a speaker to repeat the question, particularly when there is a large audience present or if the question is unclear. The presenter or the team leader should refer questions to the appropriate team members for answers. If a question is unclear, the team should seek to clarify it before trying to answer it. And as with the presentation itself, the team should practice handling questions that it thinks might arise.

There are several ways of preparing for questions while practicing talks, including:

- generate a list of questions that might arise and prepare for them;
- prepare supporting materials for points that are likely to arise (e.g., "extra" slides, computer results, statistical charts); and
- be prepared to say "I don't know," or "We didn't consider that." This is a very important point. To be caught *pretending* to know undermines the presenter's (and the team's) credibility and invites severe embarrassment.

A final note about selecting speakers is in order. Depending on the nature of the presentation and the project, the team may want to have all members speak (for example, for a course requirement), it may want to encourage less experienced members to speak in order to gain experience and confidence, or the team may want to tap its most skilled and confident members. As with so many of the presentation decisions, choosing a "batting order" will depend on the circumstances surrounding the presentation. This means that, as with all of the other matters we've touched upon, the team should carefully consider and consciously decide its speaking order.

6.3.5 Design reviews

A design review is a unique type of presentation, quite different from all the others that a design team is likely to do. It is also particularly challenging and useful to the team. As such, a few points about design reviews are worth noting.

A design review is typically a long meeting at which the team presents its design choices in detail to an audience of technical professionals who are there to assess the design, raise questions, and offer suggestions. The review is intended to be a full and frank exploration of the design, and it should expose the implications of solving the design problem at hand or even of creating new ones. A typical design review will consist of a briefing by the team on the nature of the problem being addressed, which is followed with an extensive presentation of the proposed solution. In the cases of artifacts, the team will often present an organized set of drawings or sketches that allows its audience to understand and question the team's design choices. In some cases, these materials may be provided to the attendees in advance.

A design review is often the best opportunity that the team will have to get the undivided attention of professionals about their design project. It is often scary and worrisome for the design team because its members may be asked to defend their design and answer pointed questions. A design review thus offers both a challenge and an opportunity to the team, giving it a chance to display its technical knowledge and its skills in constructive conflict. Questions and technical issues should be fully explored in a positive, frank environment. To benefit from the design review, the team should try to withhold the natural defensiveness that comes from having its work questioned and challenged. In many cases the team can answer the questions raised, but sometimes they cannot. Depending on the nature of the meeting, the team may call upon the expertise of all of the participants to suggest new ways to frame the problem or even the design itself.

Not surprisingly, such reviews can last several hours, or even a day or two. One important decision for the team is to determine, during the review, when a matter has been adequately covered and move on. This is a real challenge, since there is a natural temptation to move on quickly if the discussion suggests that a design must be changed in ways the team doesn't like. There may be a similar temptation if the team feels that the review participants have not really "heard" the team's point of view. It is important to resist both urges: Time management should not become a cover for hiding from criticism or belaboring points.

A final point about design reviews is the need to remember that conflict in the realm of ideas is generally constructive, while personality-oriented criticism is destructive. Given the heat and light that sometimes arise at design reviews, team leaders and team members (as well as the members of the audience) must continually maintain the review's focus on the design, and not on the designers.

6.4   DESIGN DRAWINGS, FABRICATION SPECIFICATIONS: DESIGNING ISN'T BUILDING

In addition to communicating with client(s) about a project, a design team must also communicate, even if only indirectly and through the client, with the maker or manufacturer of the designed artifact. This is, in a very real sense, "where the design rubber meets the road" because someone whom the design team may never meet is going to build what the team has designed. Generally, the only instructions available to the maker are those representations or descriptions of the designed object that are included in the final report. This means that these representations and descriptions must be complete, unambiguous, clear, and readily understood. What can designers do to ensure that their product descriptions will result in the built object being just what was designed?

The answer is deceptively simple: When we communicate design results to a manufacturer, we must think very carefully about the fabrication specifications we are writing. This means paying particular attention to the various kinds of drawings we do during a design project and to the different standards that we associate with final design drawings.

6.4.1   Design drawings

We first turn to design drawings, which can include sketches, freehand drawings, and computer-aided-design-and-drafting (CADD) models that extend from simple wire-frame drawings (e.g., something very much like stick figures) through elaborate solid models (e.g., elaborate "paintings" that include color and three-dimensional perspective). Drawing is very important in design, especially in mechanical design, because a great deal of information is created and transmitted in the drawing process.

In historical terms, we are talking about the process of putting "marks on paper." These marks include both sketches and drawing and *marginalia*, that is, notes written in the margins. The sketches are of objects and their associated functions, as well as related plots and graphs. The marginalia include notes in text form, lists, dimensions, and calculations. Thus, the drawings enable a parallel display of information, as they can be surrounded with adjacent notes, smaller pictures, formulas, and other pointers to ideas related to the object being drawn and designed. Putting notes next to a sketch is a powerful way to organize information, certainly more powerful than the linear, sequential arrangement imposed by the structure of sentences and paragraphs. We show an example that illustrates some of these features in Figure 6.2. This is a sketch made by a designer working on the packaging for a battery-powered computer clock. The packaging consists of a plastic envelope and electrical contacts. The designer has written down some manufacturing notes adjacent to the drawing of the spring contact. Further, it would not be unusual for the designer to have scribbled modeling notes (e.g., "model the spring as a cantilever of stiffness . . . "), or calculations (e.g., calculating the spring stiffness from the cantilever beam model), or other information relating to the unfolding design. Note that some of this information also readily translates into aspects of fabrication specifications.

Marginalia of all sorts are familiar sights to anyone who has worked in an engineering environment. We often draw pictures and surround them with text and equations. We also draw sketches in the margins of documents, to elaborate a verbal description, fortify understanding, and/or indicate more emphatically a coordinate system or sign convention. Thus, it should come as no surprise that sketches and drawings are essential to engineering

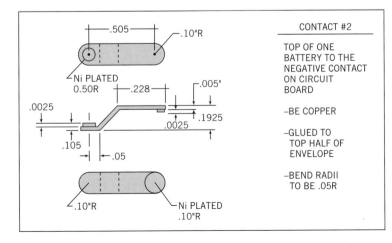

FIGURE 6.2
Design information adjacent to a sketch of the designed object. (Adapted from Ullman, Wood, and Craig, 1990.)

design. (It is interesting that while some classic engineering design textbooks stress the importance of graphical communication, drawing and graphics seem to have vanished from engineering curricula!) In some fields—for example, architecture—sketching, geometry, perspective, and visualization are acknowledged as the very underpinnings of the field.

Of particular importance to the designer is the fact that graphic images are used to communicate with other designers, the client, and the manufacturing organization. Drawings are used in the design process in several different ways, including to:

- serve as a launching pad for a brand new design;
- support the analysis of a design as it evolves;
- simulate the behavior or performance of a design;
- provide a record of the shape or geometry of a design;
- facilitate the communication of design ideas among designers;
- ensure that a design is complete (as a drawing and its associated marginalia may remind us of still-undone parts of that design); and
- communicate the final design to the manufacturing specialists.

As a result of the many uses for sketches and drawings, there are several different kinds of drawings that are formally identified in the design process. One list of the kinds of design drawings is strongly evocative of mechanical product design:

- *Layout drawings* are working drawings that show the major parts or components of a device and their relationship (see Figure 6.3). They are usually drawn to scale, do not show tolerances (see below), and are subject to change as the design process continues.
- *Detail drawings* show the individual parts or components of a device and their relationship (see Figure 6.4). These drawings must show tolerances, and must indicate materials and any special processing requirements. Detail drawings are drawn in con-

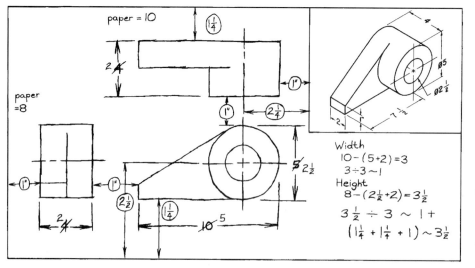

FIGURE 6.3 A *layout drawing* that has been drawn to scale, does not show tolerances, and is certainly subject to change as the design process continues. (Adapted from Boyer, et al., 1991.)

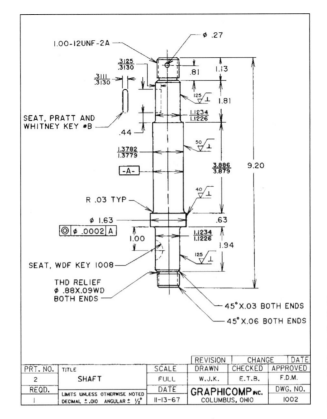

FIGURE 6.4 A *detail drawing* that includes tolerances and that indicates materials and lists special processing requirements. It was drawn in conformance with ANSI drawing standards. (Adapted from Boyer, et al., 1991.)

formance with existing standards (discussed below), and are changed only when a formal "change order" provides authorization.

- *Assembly drawings* show how the individual parts or components of a device fit together. An *exploded view* is commonly used to show such "fit" relationships (see Figure 6.5). Components are identified by a part number or an entry on an attached *bill of materials*, and they may include detail drawings if the major views cannot show all of the required information.

In describing the three principal kinds of mechanical design drawings, we have used some technical terms that need definition. First, drawings show *tolerances* when they define the permissible ranges of variation in critical or sensitive dimensions. As a practical matter, it is literally impossible to make any two objects *exactly* the same. They may appear the same because of the limits of our ability to distinguish differences at extremely small or fine resolution. However, when we are producing many copies of the same thing, we want them to function pretty much the same way, so we must limit, as best we can, any variation from their ideally designed form. That's why we impose tolerances that prescribe limits on the manufacturer and what he produces.

We have also noted the existence of drawing standards. *Standards* explicitly articulate the best current engineering practices in routine or common design situations. Thus,

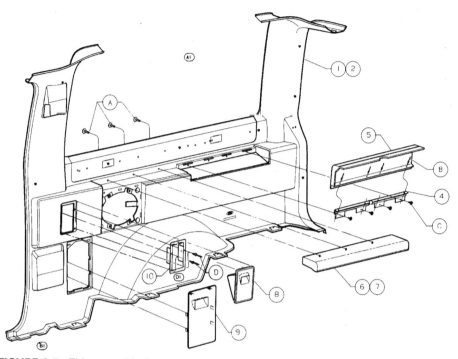

FIGURE 6.5   This *assembly drawing* uses an *exploded view* to show how some of the individual parts of an automobile fit together. Components are identified by a part number or an entry on an attached bill of materials (not shown here). (Adapted from Boyer, et al. 1991.)

standards indicate performance bars that must be met for drawings (e.g, ANSI Y14.5M–1994 *Dimensions and Tolerancing*), for the fire safety of buildings built within the United States (e.g., the *Life Safety Code* of the National Fire Protection Association), for boilers (e.g., the ASME *Pressure Vessel Code*), and so on. The American National Standards Institute (ANSI) serves as a clearinghouse for the individual standards written by professional societies (e.g., ASME, IEEE) and associations (e.g., NFPA, AISC) that govern various phases of design. ANSI also serves as the national spokesperson for the United States in working with other countries and groups of countries (e.g., the European Union) to ensure compatibility and consistency wherever possible. A complete listing of U.S. product standards can be found in the *Product Standards Index*.

6.4.2   Fabrication specifications

As we noted in Chapter 1, the endpoint of a successful design project is the set of plans that form the basis on which the designed artifact will be built. It is not enough to say that this set of plans, which we have identified as the fabrication specifications, and which includes the final design drawings, must be clear, well organized, neat, and orderly. There are some very specific properties we want the fabrication specifications to have, namely, they should be *unambiguous* (i.e., the role and place of each and every component and part must be unmistakable); *complete* (i.e., comprehensive and entire in their scope); and *transparent* (i.e., readily understood by the manufacturer or fabricator).

We require fabrication specifications to have these characteristics because we want to make it possible for the designed artifact to be built by someone totally unconnected to the designer and the design process. Further, the artifact must perform just as the designer intended because the designer may not be around to catch errors or make suggestions. The maker cannot turn around to seek clarification or ask on-the-spot questions. We have long passed the time when designers were also craftsmen who made what they designed. As a result, we can no longer allow designers much latitude or shorthand in specifying their design work because they are unlikely to be involved in the actual manufacture of the design.

Fabrication specifications are normally proposed and written in the detailed design stage (see Figure 2.4). Since our primary focus is conceptual design, we will not discuss fabrication specifications in depth. However, there are some aspects that are worth anticipating even early in the design process. One is that many of the components and parts that will be specified are likely to be purchased from vendors, such as automobile springs, O-rings, DRAM chips, and so on. This means that a great deal of detailed, disciplinary knowledge comes into play. This detailed knowledge is often critically important to the lives of a design and its users. For example, many well-known catastrophic failures have resulted from inappropriate parts being specified, including the Hyatt Regency walkway connections, the Challenger O-rings, and the roof bracing of the Hartford Coliseum. The devil really is in the details!

Of course, it sometimes is the manufacture or use of a device that exposes deficiencies that were not anticipated in the original design. That is, the way that designed objects are used and maintained produces results that were not foreseen. The F–104 fighter plane, for example, was called "the widowmaker" because test pilots found that they could do flight maneuvers that the plane's designers did not anticipate (and they also didn't think were appropriate, when they finally learned of these pilot maneuvers!). An American Airlines DC–10 crashed in 1979 because its owners did a maintenance procedure in a way that undermined the design of the engine's supporting structures and their connections to the plane's wings. How much fortune-telling ability must a designer have? How far into the future, and how well, must a designer foresee the uses and misuses to which her work will be put? There are clearly ethical and legal issues here, but for the moment our intent is only to convey the fact that design details, such as fabrication specifications, are of paramount importance.

Given that many parts and components can be bought from catalogues, while others are made anew, what sort of information must the designer include in a fabrication specification? Briefly, there are many kinds of requirements that can be specified in a fabrication specification, some of which are:

- the physical dimensions
- the kinds of materials to be used
- unusual assembly conditions (e.g, bridge construction scaffolding)
- operating conditions (in the anticipated use environment)
- operating parameters (defining the artifact's response and behavior)
- maintenance and lifecycle requirements

- reliability requirements
- packaging requirements
- shipping requirements
- external markings, especially usage and warning labels
- unusual or special needs (e.g., must use synthetic motor oil)

This relatively short list of the different kinds of issues that must be addressed in a fabrication specification makes the point about our requirements for the properties of such a specification. The specification of the kind of spring action we see in a nail clipper might not seem a big deal, but the springs in the landing structure of a commercial aircraft had better be specified very carefully!

One final note here. In the same way that there are different ways to write design specifications (see Section 1.2 and Chapter 4), we can anticipate different ways of writing fabrication specifications. We might simply specify a particular part and its number in a vendor's catalogue, which would be a *prescriptive specification*. We might specify a class of devices that do certain things, which would be a *procedural specification*. We might simply leave it up to a supplier or the fabricator to insert something that achieves a certain function to a specified level, which would be a *performance specification*.

6.4.3   Philosophical notes on specifications and drawings and pictures

Since there are so many standards that define practices in so many different engineering disciplines and domains, and these are less likely to come into play in conceptual design, we close our discussion of design drawings and fabrication specifications with a few philosophical notes.

First, different engineering disciplines use different approaches that arise in large part because of the different ways in which the disciplines have grown and evolved. These approaches continue because of the various needs of each discipline. In mechanical design, for example, in order to make a complex piece that has a large number of components that fit together under extremely tight tolerances, there is no way to complete that design other than constructing the sequence of drawings described earlier. There is no matching topological equivalent, that is, we cannot usually specify a mechanical device well enough to make it by drawing springs, masses, dampers (or dashpots), pistons, etc. We have to draw explicit depictions of the actual devices. In circuit design, on the other hand, both practice and technology have merged to the point that a circuit designer may be finished when she has drawn a circuit diagram, the analogy of the spring-mass-damper sketch. We won't discuss the many reasons that practices differ so much among the engineering disciplines, or the various practices themselves. Nevertheless, it is important that designers be aware that while there are habits and styles of thought that are common to the design enterprise, there are practices and standards that are unique to each discipline, and it is the designer's responsibility to learn and use them wisely.

We also want to reinforce the theme that some external, pictorial representation, in whatever medium, is absolutely essential for the successful completion of all but the most trivial of designs. Think of how often we pick up a pencil or a piece of chalk to sketch something as we explain it, whether to other designers, students, teachers, and so on. Perhaps this occurs more often in mechanical or structural design because the corresponding artifacts quite often have forms and topologies that make their functions rather evident. Think, for example, of such mechanical devices as gears, levers, and pulleys. Think, too, of beams, columns, arches, and dams. This evocation of function through form is not always clear. Sometimes we use more abstract drawings to show functional verisimilitude without the detail of sketches that are based on physical forms. Three examples of this sort of drawing abstraction reflect the kinds of discipline-dependent differences we discussed above: (1) the use of circuit diagrams to represent electronic devices; (2) the use of flowcharts to represent chemical-engineering-process plant designs; and (3) the use of block diagrams (and their corresponding algebras) to represent control systems. These pictures and charts and diagrams, with all of the different levels of abstraction we have seen, only extend our limited abilities, as humans, to flesh out complicated pictures that exist solely within our minds.

Perhaps this is no more than a reflection of a more accurate translation of a favorite Chinese proverb: "One showing is worth a hundred sayings." It may also reflect a German proverb, "The eyes believe themselves; the ears believe other people." In fact, a good sketch or rendering can be very persuasive, especially when a design concept is new or controversial. Drawings serve as excellent means of grouping information because their nature allows us (on a pad, at the board, and soon in CADD programs) to put additional information about an object in an area adjacent to its "home" in a drawing. This can be done for the design of a complex object as a whole, or on a more localized, part-by-part basis. Again, drawings and diagrams are very effective at making geometrical and topological information very explicit. However, we do have to remember that drawings and pictures are limited in their ability to express the ordering of information, either in a chain of logic or in time.

Our final observation in this regard is that we have not made any reference to photographic images. Certainly, photos have much of the content and impact that we ascribed to other graphical descriptions, but they are not widely used in engineering design. One possible exception is the use of optical lithographic techniques to lay out very large scale integrated (VLSI) circuits, wherein a photography-like process is used. It is also the case that we are increasingly collecting data by photographic means (e.g., geographic data obtained from satellites). With computer-based scanning and enhancement techniques, we should expect that design information will be represented and used in this way. One sign of this trend is the increasing interest in geographic information systems (GIS), which are highly specialized database systems designed to manage and display information referenced to global geographic coordinates. It is easy to envision satellite photos used together with GIS and other computer-based design tools in design projects involving large distances and spaces (e.g., hazardous-waste disposal sites and urban transportation systems). Thus, we should not forget photography as a form of graphical representation of design knowledge, alongside sketches, drawings, and diagrams.

6.5 FINAL REPORT ELEMENTS FOR THE XELA-AID DESIGN PROJECT

As required in most design projects, the student teams responsible for the chicken coop designs reported their results in the form of final reports and oral presentations. In this section we look briefly at some of the intermediate work products associated with their reports to gain further insight into some of the "do's and don'ts" discussed in Section 6.2. We will also report the conceptual design that was ultimately selected and built in San Martin Chiquito.

6.5.1 Rough outlines of two project reports

The two teams we have followed each prepared a rough outline as a first step in laying out the report structure. Table 6.1 shows the rough outline of one of the teams; Table 6.2 shows the rough outline for the other team.

These two outlines display both similarities and differences. The first team, for example, dedicated several sections to justifying their final design, while the second team organized around process. Both teams relegated sketches and drawings to appendices, although the second team put building instructions in the report body. This reflects the freedom that teams have to decide on an appropriate structure to convey their design results. This freedom, however, does not excuse them from having a logical ordering that allows the reader to understand the nature of the problem or the benefits of their solution.

As a second point on the structure of these final reports, note just how much of each could have been written during the course of the project. Each team used the formal design tools discussed in previous chapters to document its decision process. Thus, the teams could—and should—have tracked and organized their outcomes in order to facilitate the writing of their final reports.

TABLE 6.1 Rough outline for one of the chicken coop teams. The rough outline should show the overall structure of the report in a way that allows team members to divide up work with little or no unintended duplication. The structure should also proceed in a clear and logical manner. Does it for this report?

I. Introduction
II. Description of what needed to be accomplished
   A. weighted objectives tree
   B. weighted objectives tree justifications
III. Generation and evaluation of alternatives
   A. morphological chart
   B. metrics chart
IV. Design results
V. Justification of selection of coop attributes: Evaluation of our solution
VI. Project management
   A. work breakdown structure
   B. schedule
   C. budgeting
VII. Conclusions
   A. Results of our analysis of the problem
   B. Insights for next time
   C. Chicken coop suggestions
VIII. References and endnotes
Appendix: Price analysis in American dollars
Appendix: Fabrication specifications

TABLE 6.2 Rough outline for another of the chicken coop teams. This rough outline also show the overall structure; the report clearly is focused on reporting the process by which the design was arrived at as well as the actual outcome. This puts the team at some risk of writing a "history of the project" unless great care is exercised in writing the draft. This can be handled easily in the topic sentence outline.

I. Introduction
II. Client statement
III. Design process:
    A. objectives tree
    B. weighted objectives tree
    C. functional analysis
    D. morphological chart
    E. metrics and testing
IV. Final design selection
V. Building
VI. References
Appendix: Research on poultry housing
Appendix: Worksheet used to rank objectives
Appendix: Final design diagrams

Finally, neither outline will adequately translate directly into a report. There are issues that could be considered in more than one section, and others are not covered at all. Unless a team continues with a TSO or some other detailed plan, its first draft of the final report will need an unnecessarily high degree of editing.

6.5.2 One TSO for the Xela-Aid Project

Table 6.3 shows an excerpt from the topic sentence outline prepared by one of the student design teams. The overall outline was eight pages long, single-spaced. Notice that while each entry is not in itself a complete sentence, the specific point of that entry is easy to see. At this level of detail, it is relatively easy to identify redundant or inadequately covered points.

The TSO enables the team to see what will be covered within each section and each paragraph of the report. It also permits team members to take issue with or make suggestions about a section before writing and "wordsmithing" efforts are invested. For example, the team's definitions of metrics are not clear and could be challenged by a nit-picking reader (such as a professor or a technical manager). It is also unclear that all the ideas conveyed in some of the paragraphs couldn't be better covered by separating them into two paragraphs. For example, the definition of a morph chart, as opposed to the display of the team's own morph chart, might be better described separately.

Notice also that the team has adopted a historical approach to the process, which is very much at odds with recommended style. For example, "once our chart was complete we began . . ." is a red flag that the team is documenting the passing of time and events, not the design process.

Fortunately, because the team has invested effort in a TSO, it is relatively easy to make changes. If the team were to decide, for example, to move the information on chickens,

TABLE 6.3 An excerpt from a student team's TSO showing the general approach for moving from the rough outline toward a coherent first draft. Note a tendency to write in terms of chronology (e.g., "Now that we had...") and some loose definitions. The team could ask whether or not one paragraph is sufficient for each idea being carried. For example, there is no specific paragraph that identifies the particular morphological chart the team used.

V. Explanation and evaluation of feasible alternatives we considered
  A. To assist us in the determination of feasible alternatives we created what is known as a morphological chart
    1. A paragraph describing a morph chart and what it consists of
      a) functions
      b) solutions to functions (means)
  B. Once our chart was complete we began generating alternatives based on it
    1. A paragraph describing use of the morph chart
      a) Ruling out possibilities based on inconsistencies
      b) Starting with the functions with the most limiting solutions
      c) The production of valid design alternatives from combinations of solutions
  C. Now that we had a set of valid alternatives we developed methods for determining to what degree each alternative could meet our objectives
    1. A paragraph describing the definition of a metric
      a) A methodology designed to allow us to qualitatively or quantitatively measure an aspect of a design
      b) Each metric is designed to measure one aspect of the design
      c) Metrics are often derived from objectives
        i) Often used to determine the level at which an objective is being met
      d) Data collection is used to make determinations about which design components are the best (meet the most objectives) and thus determine the final design
  D. The information we collected on chickens, coops, predators, climate and the village conditions also influenced our design decisions.
    1. A paragraph or two describing the key points of our research . . .

coops, etc., to an appendix, this material can be easily included and accessed by reference to that section.

6.5.3 The final outcome: A chicken coop in San Martin Chiquito

Having used the chicken coop in San Martin Chiquito as a design example, and particularly as one in which there were either shortcomings or areas for improvement, it seems only fair and appropriate to describe the final outcome. Both teams had offered interesting designs, one a "chicken factory" conceptual design, the other based on a fenced-in area with a smaller, less expensive nesting and perching areas. After consulting with the client on a number of factors, including cost, ease of construction, and productivity, the fenced-in concept was adopted. During the summer of 1997, following the completion of the design, a team of four students and their faculty advisor traveled to Guatemala as part of a Xela-Aid team and built the coop, as well as a greenhouse and a weaving building.

Figures 6.6 and 6.7 show photographs of the coop under construction and very near to poultry occupancy. A number of changes and refinements from the initial design can be seen in the photographs, which is not uncommon in the transition from a conceptual design to its final implementation. The original design, for example, called for a cinder-block coop building. As a practical matter, it turned out to be cheaper and easier to build the coop of wood and corrugated metal. The conceptual design did not adequately address the issue of preventing the untreated wood in the base of the coop from rotting in the ground. However, on-site, a system of concrete pads was developed, onto which the wood was bolted. A further change

FIGURE 6.6 This photograph shows the chicken coop under construction in San Martin Chiquito. Notice the nestboxes on the left, and the perches on the right side. The women in the foreground include Amalia Vazquez, the leader of the women's cooperative, i.e., the client.

FIGURE 6.7 This photograph shows the chicken coop during the final stages immediately prior to occupancy by the poultry. The gate still needed attachment to the fence, and some cosmetic changes had not yet been made. In the background some of the construction materials for a weaving cooperative can be seen. The dogs in the foreground were removed and replaced with 30 chickens, tripling poultry and egg production.

was that fence posts were made of locally available angle-iron that was considerably easier to use than the triple rebar construction alluded to in our previous discussions.

At first glance, these and other changes made the final coop appear to be very different than what was originally proposed by the students. However, the basic concept did not change. A look at Figure 6.6 very clearly shows both the trays that make up the nesting boxes and the perching posts.

Today the women's cooperative in San Martin Chiquito boasts a constant production of 30 chickens and the associated egg production, a tremendous improvement as a consequence of a student design project.

6.6 MANAGING THE PROJECT ENDGAME

In this section we present some final remarks on managing the documentation activities and on closing out the project. Of particular importance to teams that want to use their experience on one project as a basis for improving their performance on the next project is a discussion of the post-project audit.

6.6.1 Team writing is a dynamic event

Most of us have considerable experience in writing papers by ourselves. This may include school term papers, technical memoranda, lab write-ups, and creative writing or journalistic experiences. Documenting a design in a team setting is, however, a fundamentally different activity than writing a paper alone. These differences turn on our dependence on co-authors, the technical demands, and the need to ensure a uniform style.

When writing as a team, we can only be sure what others are writing if all of our writing assignments and their associated content are explicit. Thus, even if any one team member's personal writing style does not include detailed outlines and multiple drafts, *all members* should do outlines and rough drafts because they are essential to the team's success. This makes issues of responsibility and cooperation important in a way that is unique for most writers. In Chapter 7 we will introduce the linear responsibility chart (LRC) as a means to ensure that work is allocated fairly and productively. The LRC should be revised and updated as part of the documentation phase.

Even with a fair allocation of work, the ultimate quality of the final report, an oral presentation, and other forms of documentation will reflect on the team as a whole and each member. It is important, therefore, to allow sufficient time for each team member to read report drafts careful. Equally important, the team must create an atmosphere in which the comments and suggestions of others are treated with respect and consideration. No one on the team should be exempt from reading the final report drafts; no one's views should be "dissed." Given the pressure under which final deliverables are often prepared, the atmosphere is, in some sense, a test of the group's culture and attitudes. The interpersonal dynamics should be monitored closely and managed carefully.

As we noted in Section 6.1, the team must also agree on a single voice. This is often quite difficult for teams, especially when several days' work is rewritten after so much effort has been expended. It is hard for any writer, and still harder for those who consider themselves skilled writers, to sublimate their styles (and egos) to satisfy a team and its designated editor. Once again, each member must keep the team's overall goals in mind.

Oral presentations also demand that a team divide work fairly. Each team member must recognize that other members may be presenting the team's work. In many cases, the presenter of a particular piece of the project may have had little to do with the element of the work being presented, or may even have opposed that approach. Once again, a central issue here is the need for mutual respect and appropriate action by the members of the team.

6.6.2   Project post-audits: Next time we will . . .

In actual practice, most projects end not with the delivery to the client, but with a *project post-audit*: An organized review of the project, including the technical work, the management practices, the work load and assignments, and the final outcomes. This is an excellent practice to develop, even for student projects or activities in which the team is disbanding completely. There is an old Kentucky saying that the second kick of the horse has no real educational value. The post-project audit is an opportunity to understand the horse better and learn where to stand next time.

The key issue in post-project audit is focus on doing an even better job next time around. As a practical matter, the post-project audit may be as simple as a meeting that takes one or two hours, or it may be part of a larger formal process that is directed by the design team's parent organization. Regardless of the scope or formal mechanism, if any, the basic post-audit process is simple:

- review the project goals;
- review the project processes, especially in terms of ordering of events;
- review the project plans, budgets, and use of resources; and
- review the outcomes.

Reviewing the project goals is particularly important for design projects, since design is a goal-oriented activity. If the project was supposed to solve problem A, then even an idea that results in earning a patent for solving problem B may not always be viewed as a success. We can only evaluate a project in terms of what it set out to do. To this end, many of the problem definition tools and techniques should be reviewed as part of the post-audit.

Closely tied to reviewing the results of using the design and management tools is the useful idea of having the team consider the effectiveness of the tools themselves. Just as a toolbox may contain many items that are only useful some of the time, many of the formal methods and techniques that are presented in this book and elsewhere will be more effective in some situations than in others. No catalog of successes or failures by the authors will have the same purchase with the team as their own experience. Reflecting on what worked and what didn't, and coming to grips with why a tool did or did not work, are both important elements of the post-project audit.

Analogously, reviewing the manner in which a team managed and controlled its work activities is also important to avoiding that "second kick of the horse." Most people learn how to organize activities, determine their sequence, assign the work, and monitor progress only through experience and practice. Such experiences are much more valuable if they are reviewed and reconsidered after the fact. As with the design tools, the management tools are not equally useful in every setting (although some, such as the work breakdown structure, appear useful in almost every situation). In commercial settings, reviewing both budgets and work assignments is critically important for planning future projects.

The last post-project audit step is a review of the outcome of the project, in terms of the goals and the processes used. While it is certainly useful to know whether or not the goals were achieved, it is important for the team members to ascertain whether this is a consequence of excess resources, good planning and execution, or simply good luck. In the long run, only teams that learn good planning and execution are likely to have repeated successes.

Our final note is that the post-project audit is not, in and of itself, a tool for assigning blame or pointing fingers. Many project and institutional settings have formal mechanisms for peer review and supervisory evaluation of team members. These can be valuable means for highlighting individual strengths, weaknesses, and contributions. They can also provide team members with important insights that they can use to improve their work in design teams. However, individual performance reviews are not central to, or even a desirable aspect of the post-project audit. The audit is intended to show what the team and organization did right to make the project successful, or what must be done differently if the project was not successful.

6.7 NOTES

Section 6.1: As noted in the text, the seven principles of technical writing are drawn from (Pearsall, 2001). In addition to Pearsall, there are a number of excellent books to support technical writing, including (Pfieffer, 2001), (Stevenson and Whitmore, 2002), and the classic (Turabian, 1996). There is no better reference to effective use of graphics than (Tufte, 2001), a classic which belongs in the library of every engineer.

Section 6.4: The list of the kinds of fabrication specifications has been adapted from (Ertas and Jones, 1996). Some of the failures mentioned, and many more, are detailed in (Schlager, 1994). Much of the discussion of drawing is drawn from (Ullman, Wood, and Craig, 1990) and (Dym, 1994). The listing of kinds of design drawings is adapted from (Ullman, 1997). The Chinese and German proverbs are from (Woodson, 1966). The drawings shown in Figure 6.2–6.4 are adapted from (Boyer, et al., 1991).

Section 6.5: The final results of the chicken coop design projects are drawn from (Connor, 1997) and (Gutierrez, 1997).

6.8 EXERCISES

6.1 How can a design team determine the audience for an oral presentation?

6.2 How does the composition of an audience affect the structure and content of a design team's presentation to that audience?

6.3 Determine the appropriate standards (e.g., ANSI standards) for presenting the design of the portable electric guitar of Exercise 3.2.

6.4 Determine whether there are any comparable appropriate standards for presenting the design of the rain forest project of Exercise 3.5.

6.5 In your role as the team leader of an engineering design team, you encounter the following situation while preparing the team's final report. Whenever Ken submits written materials documenting his work, David criticizes the writing so severely and personally that the other members of the team become quite uncomfortable. You recognize that Ken's work product does not meet your standards, but you also consider David's approach to be counter-productive. How might this conflict be resolved constructively?

6.6 You are a member of a design team that has, at last, completed its work on a design project. You have been asked by the team leader to organize a post-audit review of the team's work. Explain your strategy for conducting such an audit.

CHAPTER 7

MANAGING THE DESIGN PROCESS

You want it when?

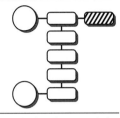

I**T SHOULD** be clear from what we have described thus far that design is an activity that can consume significant time and resources. In this chapter we explore some techniques a design team can use to manage its time and other resources. Introducing ways to manage and control a design project, we emphasize tools to successfully harness and organize all of the design team's resources—including people, time, and money—as it strives to complete a design for a client under a variety of constraints, such as limitated time and money.

7.1 MANAGING DESIGN ACTIVITIES

Much of the design environment is created by the designer, who decides, among other things, which activities are to be performed, who is going to perform them, and the order in which they are completed. Thus, creating and controlling the design environment is part of a process we term *managing design*. In this chapter we look at some tools available to the design team for planning, organizing, leading, and controlling design projects. Before we do so, however, it is useful to consider briefly some aspects of design that make design projects hard to manage. We will then see why the careful application of appropriate tools can help the designer.

A project, whether it is about design or aimed toward some other goal, can be characterized as "a one-time activity with a well-defined set of desired ends." Successful project management is usually judged in terms of scope, budget, and schedule. That is, the project must accomplish the goals (in our case a successful design), be completed within the resource limits available, and must be done "on time." Consider for a moment our beverage container example. The design must meet the concerns of the beverage company, including developing an attractive, unbreakable container that is easily and cheaply made. The designers must meet an agreed-upon budget or the design firm may not be able to stay in business over the long run. The schedule might be dictated by marketing concerns, such as producing the new container in time to sell the new juice in the coming school year. In this case, the scope of a successful design effort balances all three sets of concerns, including: achieving the client's desire to introduce a new product within the budgetary constraints of the design firm and meeting the timing dictated by the client's marketing plan. On the other hand, student design projects in a course environment have budget concerns measured in student hours, because students have other commitments, e.g., other courses and extracurricular activities. The schedule would likely reflect the timing of the course, for exam-

Project management is concerned with scope, spending, and scheduling.

ple, that the design be completed within a quarter or semester. The scope of this version of the project would address the set of client, user, and faculty concerns. In either context, design firm or design course, it is the *3Ss—scope, spending, and scheduling*—that provide the basis for the tools used to manage projects.

Having identified the 3Ss of projects, we might ask whether design projects differ from other types of projects, such as construction projects, and if so, how? It turns out that there are several important differences, the first of which lies in the definition of a project's scope. For many projects, an experienced project manager knows exactly what constitutes a success. A stadium construction project has plans that must be followed, including architectural renderings, detailed blueprints, and volumes of detailed parts and fabrication specifications. There are also accepted practices in the construction industry, so that the designation of a project manager implies that both the construction firm and the manager understand the scope of this construction project. On the other hand, the designer often cannot know what constitutes a success until a project is well underway because she may not have had a full range of discussions with the client and users. Thus, she may not be able to clarify all of the project's objectives and reconcile all the stakeholders' views. Similarly, while a construction project is expected to have only one outcome, a design project may yield many acceptable designs.

Scheduling is also different for design projects. In a construction project, the project manager can plan to do certain activities, knowing how long each will take, and can determine a logical ordering for them. For example, digging a hole for a foundation might take two weeks, but it clearly must be done before forms are built (three weeks) and concrete is poured into the hole (two days). Assembling and organizing such planning and scheduling data allows a project manager to determine how much time the project will take. In many cases, the manager of a design project asks how much time is available, rather than summing up how long the tasks take. This approach to scheduling represents the design team's intent to use all of the available time to generate and consider several viable design alternatives, while still trying to meet the constraint of finishing the project by a specified time.

These differences lead us to wonder whether project management techniques are appropriate for use in engineering design projects. After all, if the scope is ambiguous and the timing is up to the client (and may even seem arbitrary), how relevant are tools developed for well-understood projects? It turns out that the uncertainties and external impacts associated with design projects tend to make some of the management tools even more useful and necessary. As this chapter unfolds, we will see that project management tools can be useful for gaining consensus by the design team about what must be done, who will do it, and when things must be done—even if our expectations about the form of the final design are initially up in the air.

The team nature of design projects also lends support to the use of project management tools. In Chapter 2 we highlighted the stages of team formation. Many important issues must be addressed when the team moves into the performing stage, including the need to effectively communicate the activities, schedule, and progress to each team member and to other stakeholders. Work must be allocated fairly and appropriately. And we must ensure that tasks have been done properly, in a sequence that allows team members who depend on prior work to plan their actions. The management tools we introduce here are helpful in each of these circumstances.

7.2 AN OVERVIEW OF PROJECT MANAGEMENT TOOLS

Recall from Chapter 2 that we can model the process by which we eventually move from a client's problem to a detailed design of a solution. This process uses a number of formal design methods and several means of gathering and organizing information in order to generate alternatives and evaluate their effectiveness. While the methods and means discussed can be assigned to the steps in the design process, design is not a simple "cookbook" process. Similarly, managing the design process also requires more than rote application of project management tools. In this section we briefly describe the tools that we use to plan a design project, organize design activities, agree on responsibilities for the project, and monitor progress.

We noted in Chapter 1 that management consisted of four functions:

- *Planning* a project leads immediately back to the 3S model of project requirements. We must define the scope of the project, determine how much time we have to accomplish the scope (scheduling), and assess the level of resources (spending) we can apply to the project.

- *Organizing* a project consists largely of determining who is responsible for each task area or activity of the project, and which other human resources can be called upon.

- *Leading* a project means using tools to motivate a team by showing that the tasks are understandable, the division of work is fair, and the level of work produces satisfactory progress toward the team's goals. Note, however, that project leadership cannot be provided by tools alone.

- *Controlling* can only be done in the context created by the 3Ss and the plans that support them. Tracking progress is meaningful only against a stated set of goals. Further, we can change our plans or take corrective action only if the team is confident that the plans they have developed are, in fact, going to be used.

The primary tool used to determine the scope of activities is the *work breakdown structure* (WBS). The WBS is a hierarchical representation of all of the tasks that must be accomplished to complete a design project. Project managers use WBSs to determine which tasks must be done. They will generally break the work down (hence the name) into pieces sufficiently small that the resources and time needed for each task can be estimated with confidence.

The *linear responsibility chart* (LRC) identifies which team member has the primary responsibility for the successful completion of each task in the WBS, and identifies others who must participate in finishing that task. The LRC uses a matrix format to match each of the tasks requiring management responsibility with the members of the team, the client, users, and other stakeholders. This is particularly important for team-based activities, both to clearly identify who is responsible for each task and to indicate all individuals that should be involved (e.g., teammates, the client, or an external expert).

We can schedule activities in several ways:

- A *team calendar* shows all of the time that is available to the design team, with highlights that indicate deadlines and time frames within which work must be completed.

- A *Gantt chart* is a horizontal bar graph that maps various design activities against a time line.
- An *activity network* graphs the activities and events of the project and shows the logical ordering in which they must be performed.

It is important to be aware that while a schedule can provide vital assistance for planning and controlling the work of a design team, it can easily be little more than a flashy picture or marketing graphic.

A *budget* is a list of all of the items that will incur an economic cost, organized into some set of logically related categories (e.g., labor, materials, etc.). The budget is the key tool for managing spending activities in a project. Note that there is an important distinction between the budget for doing the design, or design activities, and the budget needed to produce or build the artifact being designed. Our concern is primarily with the budget needed to do a design.

Other control methods are used to manage projects, but many are simply not well suited to design projects. For example, *earned value analysis* relates costs and schedules to planned and completed work. While earned value analysis is useful for certain large-scale projects with effective reporting systems, it is overkill for the smaller, team-based projects that we discuss. On the other hand, the *percent-complete matrix* (PCM) relates the extent of the work done to the total level of all work to be done and is a more useful and appropriate tool. We will develop a version of the PCM that is appropriate for smaller team design activities.

In the following sections we discuss and provide examples of the tools mentioned above. Just as each of the design methods and means are not be appropriate for every design project, these management tools will not have to be used by every team on every project. However, since the tools are important to good, team-based design work, and we can never predict with certainty the kinds of design activities we will undertake in the future, these management tools are worth having in our individual arsenals.

7.3 WORK BREAKDOWN STRUCTURES: WHAT MUST BE DONE TO FINISH THE JOB

Most of us would be a bit overwhelmed if asked to describe exactly how to start and drive a car, even though this is a common task. We might start with something like "first you get in the front seat"—which presupposes that you know how to get into the car! If forced to describe this task to a someone from a country with few cars and limited English-language skills, we might want to break down the task into a number of task groups, such as getting into the car, adjusting the seat and mirrors, starting the engine, driving the car, and stopping the car. We might even want to review the entire plan before starting the car so that a driving student would know how to stop before starting out. This decomposition of tasks or concepts is the central idea of the work breakdown structure (WBS). When we are confronted with a very large or difficult task, one of the best ways to figure out a plan of attack is to break it down into smaller, more manageable subtasks.

The WBS is the most important management tool for design projects—it organizes all of a project's tasks.

A more compelling example is that of a team that has been asked to design a spacecraft. The team will have to design across several specialties, including propulsion, communications, instrumentation, and structures. Here the design team leader will work very hard to ensure that the team's propulsion experts are actually assigned to propulsion tasks, so that the experts work on tasks relevant to their expertise. In order to do this properly, the team leader must determine just what those tasks are. The WBS is a listing of all the tasks needed to complete the project, organized in a way that helps the project leader and the design team understand how all of the tasks fit into the overall design project.

We show a work breakdown structure for the beverage container design example in Figure 7.1. At the top level it is organized in terms of eight basic task areas:

- Understand customer requirements
- Analyze function requirements
- Generate alternatives
- Evaluate alternatives
- Select among alternatives
- Document the design process
- Manage the project
- Detailed design

We see also that each of these top level tasks can broken down in greater detail. Because of page size limitations, in this example we show great detail only for some of the tasks (e.g., understanding customer requirements). If we were actually part of a team carrying out this project, we would go into much greater depth in all of the areas. Also, note that this method of organizing the work is not the only way to structure a WBS. We will see several alternative organizing frameworks later in this chapter.

Several observations about the WBS depicted in Figure 7.1 are in order. First, the basic principle for a WBS is that each item that is taken to a lower level is *always* broken down into *two or more subtasks* at the lower level. If the task is not broken down (so the lower level is a lone entry), then either the lower level is incomplete, or it is just a synonym for the upper level. Second, if we cannot determine how long an activity will take or who will do that activity, then the key WBS rule dictates we should break it down further. In fact, experienced project managers will be more inclined to have shorter, less detailed WBSs than relatively inexperienced managers because their greater experience makes them able to aggregate subtasks into identifiable, measurable tasks.

Our third observation is that a WBS should be *complete* in the sense that any task or activity that consumes resources or takes time should be included either in the WBS explicitly or as a known component of another task. That is why the tasks of documentation and management are shown in Figure 7.1. Activities such as writing reports, attending meetings, and presenting results are essential to the completion of the project, and failure to plan for them will certainly result in problems later. Estimating who and what is required, and for how long, is a valuable discipline for any project, whether in a design course or in the "real world," both for developing the design and for ensuring that there is sufficient time to document and present results.

Our final observation on the WBS is that any part of its hierarchy of tasks should *add up*, that is, the time needed to complete an activity at a top level should be the sum of the times for tasks listed at the level below. Thus, breaking work down to the next level below must be done thoroughly and completely.

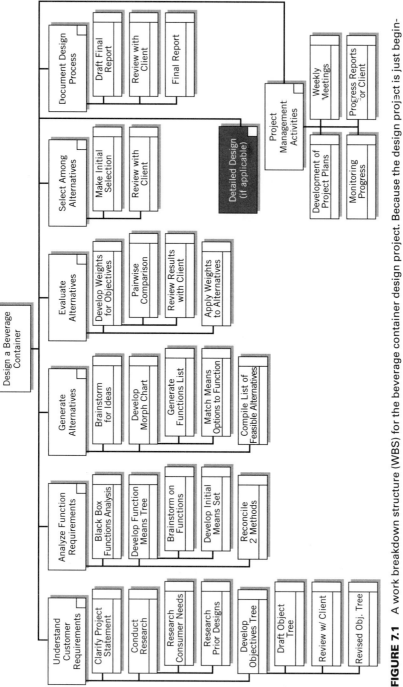

FIGURE 7.1 A work breakdown structure (WBS) for the beverage container design project. Because the design project is just beginning, the structure necessarily takes on a formal and somewhat generic framework. Note, however, that the designers are already aware of some details, such as the distinction between identifying consumer needs and prior designs.

The last two observations about WBSs provide us with two criteria for evaluating their utility:

- *Completeness* means that the WBS must account for all of the activities that consume resources or take time.
- *Adequacy* means that tasks should be broken down to an adequate level of detail, such that the project team can determine how much time it will take to do them.

It is also important for us to note what the WBS is *not*. First, a WBS is *not* an organization chart for completing a project. This may be confusing because "org charts" are visually similar graphics. The WBS is a break down of tasks, *not* of titles, roles, or people in an organization. Second, a WBS is *not* a flow chart showing temporal or logical relationships among tasks. In many cases the listing of the tasks will be organized in such a way that a task (e.g., writing the final report) is shown in a different part of the hierarchy than other tasks which must precede it (e.g., all of the design, building, and testing that is being reported). Third and last, a WBS is *not* a listing of all the disciplines or skills that are required to complete the tasks. In many cases the tasks to be completed may require a number of different skills (e.g., electrical engineering and propulsion engineering). Tasks

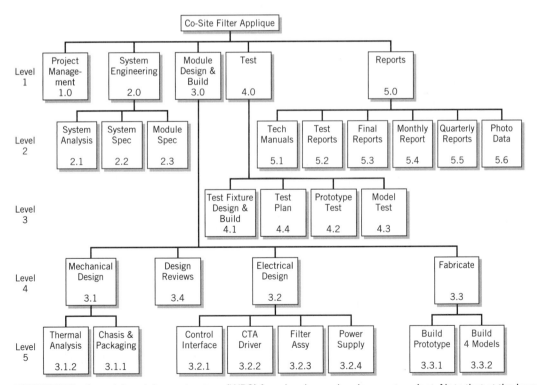

FIGURE 7.2 A work breakdown structure (WBS) for a hardware development project. Note that at the lowest level (Level 5), the electrical design tasks have been broken down to the level of the components to be designed. At this stage, the designer of the Power Supply might follow another WBS similar to Figure 7.1. (Adapted from Kezsbom, Schilling, and Edward, 1989.)

performed by professionals with these skills can be combined into the same part of the hierarchy if that listing of the tasks meets the above criteria of completeness and adequacy.

Figure 7.2 shows another sample WBS, in this case taken from an electrical hardware project. Here the design subtasks are organized in terms of electrical and mechanical design. This may appear to violate our concern with disciplines noted above—until we realize for this project this is simply a convenient way to organize *the tasks*, which is always our key intent for a WBS.

Figure 7.3 shows yet another sample WBS, built in a software package, Primavera Project Planner, for the engineering tasks of an car company. Note that this WBS is not even in a graphical form, although it is hierarchical. While graphical forms may offer clarity, it is perfectly permissible to use a tabular form such as that shown in Figure 7.3. In fact, tables are a common means of collecting the information together once a WBS has been developed into its final form. Note also that this example has the work broken down by the vari-

---

PRIMAVERA PROJECT PLANNER

Date 08JAN98   -----WORK BREAKDOWN STRUCTURE-----

ENGR - Active Projects for the Fiscal Year

Structure : xxx.xxx.xx.x

WBS Code   Title

94 All Projects
   94E All Engineering Projects
      94E.101 Project E101
         94 E.101.A General
         94E.101.A7
         94E.101.B Air Bag
         94E.101.C Mechanical Release System
         94E.101.D Electrical Systems
         94E.101.E Interior Dashboard
         94E.101.F Structural Door System
   94E.102 Retrofit Automobile Plant
         94E.102.A Enclosure
         94E.102.B Structural System
         94E.102.C Mechanical System
         94E.102.D Electrical System
         94E.102.E Estimating
         94E.102.F Specifications
         94E.102.G General
   94I All Installation Projects
      94I.101 Tooling & Equipment Installation
         94I.101.A Structural Slab
         94I.101.B Piping
         94I.101.C Equipment
         94I.101.D Electricity
         94I.101.E Interior Finishes
         94I.101.F Ventilation & Plumbing
         94I.101.G General

FIGURE 7.3

A work breakdown structure (WBS) for the engineering projects of an automotive firm. This nongraphical WBS organizes the firm's activities according to systems for the autos and overall factory installation projects. The level of detail is not very high, and presumably the firm would have supporting WBSs for some if not all these projects.

ous car components. This is also permissible, so long as the WBS satisfies our concerns about completeness and adequacy.

In the end, the WBS is a tool for a project team to ensure their understanding of the tasks needed to complete their project. That is why a WBS is so valuable for determining the scope the project.

7.4 LINEAR RESPONSIBILITY CHARTS: KEEPING TRACK OF WHO'S DOING WHAT

Once the tasks to be done are identified in a WBS, a design team has to determine whether or not it has the people, the human resources, to accomplish those tasks. The team also has to decide who will take responsibility for each task. This can be done by building a *linear responsibility chart* (LRC). The LRC lists the tasks to be managed and accounted for and matches them to any or all of the project participants. Figure 7.4 shows a simplified LRC corresponding to many of the tasks in the beverage container design WBS of Figure 7.1. In addition to all of the top-level tasks, the subtasks associated with several of the lower levels are also given. In practice, it is advisable to show all of the top-level tasks, as well as those subtasks that may require management attention. Because of the evolving nature of design projects, less experienced project managers may want to assign responsibility only for the top-level tasks and the next set of tasks that require the team's attention. This allows the team's roles and responsibilities to develop with experience, along with the project.

As we can see in Figure 7.4, there is a row for each task, within which the role, if any, of each project participant is given. These roles do not necessarily mean assuming the primary responsibility. Indeed, most of the participants will play some sort of supporting role for many of the tasks, including reviewing, consulting, or working at the direction of whomever is responsible. A column is assigned to each of the participants; this allows them to scan down the chart to determine their responsibilities to the project. For example, we see that the client (or the client's liaison) will be called upon to give final approval to the objectives tree, the test protocol, the selected design, and the final report. The liaison must also be consulted during some of the pre-design activities, and will be asked to review various intermediate work products. The client's research director is somewhat of a resource to the team, and may be consulted at different points of the project, but *must* be consulted regarding the test protocol. The team's boss, the director of design, has asserted the right to be kept informed at a number of points in the project, most notably in terms of design reviews. The project also has access to one or more outside experts, who are generally available for consultation, and who must review the design and certain other (unspecified) documents.

We also note from the LRC in Figure 7.4 that the team leader does not always have the primary responsibility for the project. It is often the case in team projects that the team leader will not be responsible for tasks that are outside of her technical area of expertise, although she may want to specify a review or support role in order to remain informed. Sharing responsibility this way is sometimes quite difficult both for team leaders and teams. Sharing responsibility, which is strongly tied to the storming phase of group formation, requires practice. Therefore, the LRC can be used to make this part of team formation more explicit to the team, and to allow the team to reach consensus on who will be doing what in the project.

LRCs can ensure that the workload is distributed fairly and equitably.

Linear Responsibility Chart	Team Member #1	Team Member #2	Team Member #3	Team Member #4	Team Member #5	Director of Design	Client Liaison	Client Research Director	Outside Consultant
1.0 Understand Customer Requirements	1								
1.1 Clarify Problem Statement	1	2	2	2	2		3	4	
1.2 Conduct Research	1	2		2	2		4	4	4
1.3 Develop Objectives Tree	1								4
1.3.1 Draft Objectives Tree			2	2		5	5	3	4
1.3.2 Review w/ Client	1		2			5	5	3	4
1.3.3 Revise Objectives Tree	1		2	2		6		4	
2.0 Analyze Function Requirements	2	2	1	2	2	5	4	3	3
3.0 Generate Alternatives				1					
4.0 Evaluate Alternatives	5	1	2	2	2				
4.1 Weigh Objectives	1	2				5	6		
4.2 Develop Test Protocol	5	1			2	5	4	3	3
4.3 Conduct Tests		1	2		2			5	3
4.4 Report Test Results	5	2	2		1	5	5	5	5
5.0 Select Preferred Design	1	2			2	5	6	4	4
6.0 Document Design Results		1							
6.1 Design Specifications	1			2		6			
6.2 Draft Final Report	5	1		2		5	5		4

Linear Responsibility Chart	Team Member #1	Team Member #2	Team Member #3	Team Member #4	Team Member #5	Director of Design	Client Liaison	Client Research Director	Outside Consultant
6.3 Design Review w/ Client	1	2		2		5	3	4	3
6.4 Final Report	5	1		2	2	5	6	4	4
7.0 Project Management	1								
7.1 Weekly Meetings	1	2	2	2	2				
7.2 Develop Project Plan	1	2	2	2					
7.3 Track Progress	1					5			
7.4 Progress Reports	1						5		

Key:

1 = *Primary responsibility*

2 = *Support/work*

3 = *Must be consulted*

4 = *May be consulted*

5 = *Review*

6 = *Final Approval*

FIGURE 7.4 A linear responsibility chart (LRC) for the beverage container design project. Each participant in the project can read down his column and determine his responsibilities over the entire project. Alternatively, the Project Manager can read across a row and determine who is involved with each task.

The LRC can also be used to let outside stakeholders in a project understand what they are expected to do. In the beverage container example, the client's research director clearly has an important role to play in the safe conduct of the testing phase. It is very important this person knows early on what is expected, and be allowed to plan accordingly. Similarly, the outside experts may need to allocate time to insure availability, and the director of design may need to make resources available to pay for the experts' time.

It should be clear by now that the LRC can be a very important document for translating the "what" of the WBS into the "who" of responsibility. At the same time, we might be tempted to use the LRC to avoid admitting that the team doesn't know something. For example, if every task has every team member assigned in a support or work role, it should raise serious doubts in our minds about whether we really do understand those roles. Similarly, if a team leader claims primary responsibility for all the tasks, her team will certainly be tempted to consider the LRC as little more than a power grab or a mirror of the team leader's insecurities. Thus, it is better to consider a matter open and leave its row blank than to fill it in blindly for the sake of a (false) conclusion. It is also important that a team understand that it may be necessary to revisit roles as the project unfolds, especially if the team is relatively inexperienced or the project is initially ambiguous.

7.5   SCHEDULES AND OTHER TIME MANAGEMENT TOOLS: KEEPING TRACK OF TIME

Scheduling and similar time management tools help us identify in advance those things that will really mess up our project if they aren't done on time. Three primary scheduling tools are frequently used in project management: a calendar, an activity network, and a Gantt chart. A team calendar is likely the most familiar tool as it performs many of the same functions as personal calendars or diaries: It maps project deadlines or due dates onto a conventional calendar.

The activity network and the Gantt chart are more powerful and, consequently, possibly more useful. Both are graphical representations of the logical relationships between tasks and the time frames in which they are to be done. Indeed, most software programs for project management use the same information to generate both activity networks and Gantt charts. However, there are important differences in practice that make it worthwhile to describe both so that a team can decide which tool is most appropriate for its use.

7.5.1   Team calendars: When are things due?

As we have just said, a team calendar is simply a mapping of deadlines onto a conventional desk or wall calendar. Such deadlines will certainly include externally imposed ones, such as commitments to clients (or to professors for academic projects), but should also include team-generated deadlines for the tasks developed in the WBS. In this sense, the team calendar is really an agreement by the team to assign the resources and time necessary to meet the deadlines shown on the calendar. Figure 7.5 shows a team calendar for a student design team seeking to complete its project by the end of April, an externally imposed deadline.

FIGURE 7.5 A team calendar for a student design project. Note that externally imposed deadlines, team commitments, and recurring meetings are all included on the calendar. It is usually better to make the team calendar "too complete" than it is to leave out potentially important milestones or deadlines.

Note that the calendar includes several deadlines over which the team probably has no control, such as when the final report is due and when in-class presentation of results is to be done. It also includes routine or recurring activities, such as Tuesday night team meetings. Finally, it includes some deadlines that the team has committed to realizing, such as completing a prototype by 5 PM on April 2.

Team calendars should be reviewed weekly (at least!).

Several points should be kept in mind in setting up a team calendar. First, the idea of a team calendar implies that the deadlines are all understood and agreed to by everyone on the team. As such, the calendar becomes a document that can—and should—be reviewed at every team meeting. Second, the team calendar should allow times that are consistent with the time estimates generated in the WBS. If a task was determined to take two weeks to complete, there is little point in allowing that task only one week on the team calendar. A final point to note is that the team calendar, while easily understood by members of the team, *cannot by itself capture the relationship between activities*. For example, in Figure 7.5 we see that building the prototype precedes proof of concept testing *only* because the team chose to put it that way. For many artifacts a proof of concept may actually precede building a final prototype. The team calendar cannot address this sort of problem, nor can it "remember" team decisions of this sort. For this reason, a team calendar is most useful for small projects or in cases where it is supplemented with other project management tools (such as those we discuss in the following sections).

7.5.2 Activity networks: Which tasks must be done first?

An activity network is based on the idea that each task can be treated as a separate activity, and its output or result as an event. Recall our earlier example of the construction project. "Digging a hole for the foundation" can be considered an activity, and the "existence of the foundation hole" as an event. We could also consider the task "build forms" as an activity, and "forms erected" as an event. Finally, we could consider the task "pour concrete into forms" as another activity, and "foundation completed" as yet another event. Now, there are at least two ways that we could represent this set of activities and events graphically. For example, we could construct a network of nodes and connecting arcs, and then identify each node with an event and each arc as the activity that caused the event to happen. Alternatively, we could represent each activity with a node, and identify the events simply as the arcs that extend from the nodes. It makes no logical difference which scheme we choose, *as long as we are consistent*. We will use the form of activity network called the *Activity-on-Node* (AON) network, which adheres to the convention of placing activities on or at nodes. This form of activity network is the one most often used in current project management software.

If we were to take a list of all of our tasks, such as that generated by a very complete WBS, we could create a list or set of nodes, one for each activity. However, because we are interested in scheduling activities, it is absolutely essential that we properly determine the logical ordering, or *precedence relationships*, between them. In the simple construction example mentioned above, only a very foolish (or dangerous) construction project manager would expect to pour the concrete before the forms were constructed and in place. Similar relationships often exist in design projects. We would not, for example, attempt to determine which of a user's objectives are most important until we were sure that we had fully enumerated all of the objectives. Similarly, a team leader should encourage the team to defer evaluating alternatives until after all of the alternatives have been generated. We thus see in these cases that the logical ordering of the project's activities will keep us from doing one activity until another is complete. This type of logical relationship is called *finish-to-start precedence*. That is, we must finish one activity prior to starting its successor activity.

Sometimes the logical relationships are quite subtle. We might, for example, get a potentially useful design idea while we are still learning about the client's needs. In this case, the relationship is such that we can *begin* the task of generating ideas before we complete the task of understanding needs, but we can't consider the idea-generating task *complete* until we are quite sure that we have finished the task of understanding needs. This is called *finish-to-finish precedence*, and it means that we cannot finish the successor until we are sure that the prior activity is finished. One common example of finish-to-finish precedence occurs during the writing of a final report. Teams are sometimes encouraged to begin to write the final report of a project very early on, since that allows them to get sections such as the literature review done while the literature is still fresh in their minds. However, it would be very disturbing to learn that a team had finished the final report before it had completed the design aspects of their project!

Managers need to understand both logical and temporal relationships between tasks.

A third type of relationship is *start-to-start precedence*, in which one activity cannot be started until another one is started as well, although neither activity needs to be finished before the other. An example of this type of relationship might be editing sections of the final report for a project. In order to have a team member check grammar and style, some of the report must already have been written, but it doesn't have to be finished for checking to get started. (Clearly, in this and similar instances, judgment must be exercised. Checks done before report sections are substantially written may well have to be repeated because minimally completed sections are subsequently extended.) It is important to understand the different types of relationships between tasks, because if we assume that all activities are either fully independent or obey finish-to-start precedences, we are wasting opportunities to get work done earlier when resources may be more readily available or even idle.

Once we understand the logical relationships, we can draw an activity network for our project. Basically, this network consists of a node for each activity and an arrow out of each activity to its logical successors. Usually a number of activities do not have any logical predecessors or logical successors. At the start of a project, activities such as basic research or initial meetings with the client may not depend on anything before them; to deal with this, there is a convention that calls for starting an activity network with a "Start Project" node. In some cases, there may actually be a formal start, such as a project initiation meeting, but in others this is what is known as a *dummy activity*. Similarly, to ensure that every node has some place to connect its outward arrow, there is by convention an "End Project" node. Once again, there may be a formal ending to the project, such as turning in the final report or meeting with the client for the last time, but if not, the network should have a dummy node at its end. One consequence of the "Start Project" and "End Project" convention is that every node in the network except these two will have at least one arrow leading into it and at least one arrow leading out. If the activity has logical precedents, then the inward arrow should come from them. If the activity has logical successors, then the arrow(s) out of the activity should point to the successor.

We show in Figure 7.6 a simple activity network for the beverage container design project that indicates the logical relationships between the project's activities. Such a network can also take advantage of and incorporate the known duration of each activity.

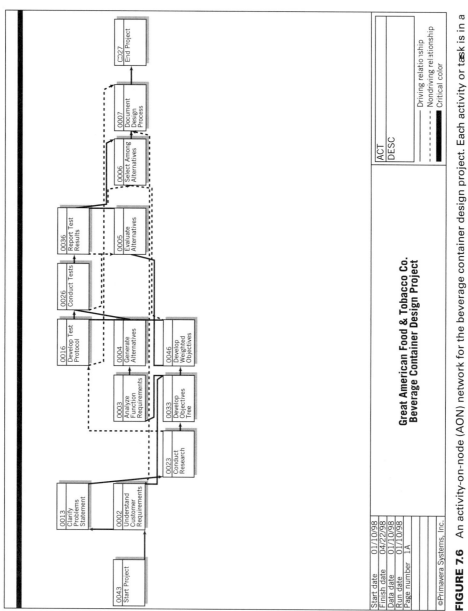

FIGURE 7.6 An activity-on-node (AON) network for the beverage container design project. Each activity or task is in a box, or node, and the logical relationships with other activities are shown using arrows. Note that the logical relationships can take several forms (start-to-start, finish-to-start, etc.).

(Recall that in Section 7.3 we said that each task in the WBS should be broken down at least to a level such that we can estimate how long it will take to complete that task.) The activity network does not, however, fully schedule the tasks. The logical relationships in Figure 7.6 tell us that we cannot conduct the testing until we have both developed the test protocol and generated alternatives (which may include building mockups or prototypes). Note, however, that we can begin to develop the test protocol as soon as our research has been completed and we have met with the client's research director. If the research takes only a few weeks, and building a testable mockup takes much longer, then we have a number of different times at which we could decide on a protocol. There is no simple formula for deciding when to decide; this decision calls for management judgment. If we think the team is going to be really busy during part of this time frame and idle during another (such as when waiting for parts delivery), we would want to plan on doing the protocol during that idle time. This illustration highlights an important notion, namely, that there are some activities whose start time we can adjust without affecting the overall time-to-completion of the project, while there may be other activities we should start as soon as logically possible. An activity for which we can adjust a start time is said to have *slack*, the time difference between the earliest date we can start that activity and the latest date we can start it without affecting the overall timing of the project. Activities that have no slack are said to be on the *critical path*. Starting and completing such critical path activities on time is of central importance to the timely completion of a project. That is why successful project managers pay very close attention to activities that are on the critical path, while staying "only" mindful of the others.

> Activities that are on the critical path must be completed on time or the entire project will be delayed.

We certainly do not intend to fully explain all of the subtleties of this type of scheduling, but it is worthwhile for a design team to determine which activities must be carefully kept on schedule, and which are off the critical path. Adjusting activities that are not on the critical path is also important for balancing the team's workload. In fact, it is generally better to plan a team's work so that it can do things at a steady or relatively constant pace, since normal occurrences of Murphy's Law always create opportunities for things to pile up near the end. Therefore, it is helpful to plan the work and move activities with some slack toward the beginning of the project. We sometimes say in design projects, it is better to organize the work so we can panic early; later events that might induce panic have already been dealt with. (Experience suggests that you will still get the chance to panic later even if you do panic early, but it likely will be about more interesting and useful problems.)

7.5.3   Gantt charts: Making the timeline easy to read

The same information that we incorporate into an activity network can also be represented in a bar graph or *Gantt chart*. This approach to scheduling is said to have originated with Henry Gantt, one of the early founders of the field that is now known as industrial engineering. Gantt was charged with improving the output of munitions production during the World War I, and he determined that it would be useful to formally track and diagram the processes involved. Apparently he was quite successful, as his method is widely used today not only to track processes, but to plan them.

Figure 7.7 shows the Gantt chart corresponding to the activity network of Figure 7.6. Note that for each activity there is a start date and an end date that can be read from the time

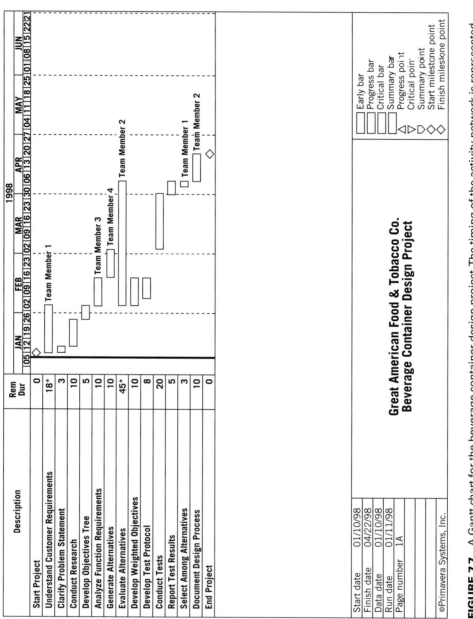

FIGURE 7.7 A Gantt chart for the beverage container design project. The timing of the activity network is represented as a bar chart. Activities marked with an asterisk in the remaining duration column are summaries of their subtasks. Notice the difficulty in determining logical relationships even in this small example.

axis at the top of the chart. Many managers find that the Gantt chart is more readable than an activity network since it allows them to relate time in a rather evident manner. Further, we can develop or sketch a Gantt chart without any computer-based tools, using only graph paper, although we can also easily do them within standard spreadsheet programs. This simplicity can come at a price, however. A project team may construct a Gantt chart without carefully thinking through the logical relationships discussed above. If this happens, then several problems will likely ensue as the project unfolds. First, it becomes harder for the team to set priorities if the logical relationships among tasks are unclear. The team may, for example, work on the most interesting or most difficult task, even if others are on critical path. In addition, in the absence of accurate information regarding the logical ordering of events, the team may find that a seemingly unimportant activity has more important successors that cannot be begun. Finally, if the inevitable slippage occurs on some tasks, all of the slack in an activity can be consumed, which then makes that activity and its successors critical.

These sorts of problems can be avoided if the activity network is assumed to relate to a Gantt chart with finish-to-start precedence. Thus, teams should work as a group to construct an activity network first. The results team can then be displayed in a Gantt chart or some other graphical format. This helps ensure that all of the tasks and their relationships to each another have been considered, and it reinforces the planning already undertaken in developing the LRC.

7.6 BUDGETS: KEEPING TRACK OF THE MONEY

Budgets are difficult but essential tools for project management. They permit teams to identify the financial and other resources required, and to match those requirements to the available resources. Budgets also require teams to account for project monies. Finally, budgets serve to formalize the support of the larger organization from which the team is drawn.

Many of the engineering economics concepts that we will discuss in Chapter 8 are relevant for budgets, so we will defer much of our discussion of these concepts at this time. Further, we usually don't need large, complex budgets for doing the sort of design projects that are likely to be done in academic or similar settings. (Remember, as we have said before, we are concerned with the *budget for doing the designs*, not with the budget for making the designed objects.) Thus, design project budgets normally include research expenses, materials for prototypes, and support expenses related to the project.

We limit our budget discussions to the above categories of costs, that is, materials, travel, and incidental expenses. This means that in attempting to budget for a design project, it is necessary at the outset to try to determine what sorts of solutions are *possible*. This is not to say that we have determined the solution, rather that we should consider resource needs earlier than might really be desirable. One effect of this is that design teams often try to establish *not to exceed* budgets that set limits on expenses by specifying the highest cost that might come about. The danger in this approach is that if it is done this way routinely, for all projects in an organization, resources may be set aside for design projects that will never used.

As a final note, it is important to properly value the time invested in a design project by each and every member of a design team. This is important even in student design projects. (In fact, there is a tendency to undervalue this very scarce resource just because we did not call out the time in the budget.) One way to place a value on a team member's time is to adapt the "algorithm" that employers use to "bill out" the time of an engineer who is working on a project. Most firms charge between *two and four times* an employee's direct compensation when they bill a client for that employee's time. That multiplier covers fringe benefits, overhead costs, supervision, and profit. If we were to bill student time at a minimum wage rate of only $8.00 per hour, a team of four students working ten hours each week on a project for ten weeks would be billed out by a design firm at $6,400–12,800 for the entire project. Put in the simplest terms, time is a valuable, scarce, and irreplaceable resource—don't waste it!

7.7 TOOLS FOR MONITORING AND CONTROLLING: MEASURING OUR PROGRESS

We have now developed a plan, a schedule, and a budget. How do we track our team's performance relative to the plan? This is an important question, but it can be very hard to answer. The manager of a construction project can go out and see if a task has been done by the date planned. In a design project, however, monitoring and control is more subtle and, in some ways, more difficult. Therefore, it is essential that the members of a team agree upon a process for monitoring their joint progress before the project gets very far underway.

There are a number of techniques and tools available to monitor projects, but these often involve team members filling out time sheets, punching clocks, or other accounting tools. For smaller design projects, especially academic projects, these kinds of tools may not be very effective. We will therefore present a simplified version of the *percent-complete matrix* (PCM) that is widely used in industry to relate the extent of work done on the parts of a project to the status of the overall project.

The goal of the PCM is to use the information in the WBS and the budget to determine the overall status of the project. Constructing a PCM requires only that we know the cost of each item or area of interest, and the percent of the total cost corresponding to that item. Then the PCM allows us to input the percent of the work on that task or work item and, by summing over all the items in the project, we can calculate the total percent of the project completed. In general, the method is best suited to cases where there some clear method of calculating progress is available. If, for example, the foundation work of a building project constitutes 25% of the total expected costs of a project, then we have completed at least 12.5% of the project when we have completed one-half of the foundation work. A manager can periodically update progress in each of the general areas in the WBS to determine overall project progress.

In some cases a physical measure can serve as a proxy for progress, such as yards of concrete poured compared to the total volume called for in the plan, or tons of steel erected

Percent Complete Matrix

Task	Planned Duration (days)	Percent of Total	Status (see key)	Credit (days)
Start Project	0	0%	2	0.0
Clarify Problem Statement	3	3%	2	3.0
Conduct Research	10	11%	2	10.0
Draft Objectives Tree	2	2%	2	2.0
Review OT	1	1%	2	1.0
Revise OT	2	2%	2	2.0
Analyze Functions	10	11%	1	3.3
Generate Alternatives	10	11%	1	3.3
Develop Weighted Objectives	10	11%	2	10.0
Develop Test Protocol	8	9%	1	2.6
Conduct Tests	20	21%	0	0.0
Report Test Results	5	5%	0	0.0
Select Among Alternatives	3	3%	0	0.0
Document Design Process	10	11%	0	0.0
End Project	0	0%	0	0.0
Total Days Budgeted	94	100%		39.6%
Key: 0 = Not Started, No Credit, 1 = In Process 1/3 Credit, 2 = Completed, Full Credit				

FIGURE 7.8 A percent complete matrix (PCM) for the beverage container design project. Each activity and its share of the overall project is given. The team is given 33 percent credit when an activity is begun and the balance upon completion. Unless the tasks have been broken down sufficiently, this method can be misleading, but for small projects it provides a reasonable approximation of progress.

compared to budgeted totals. While this approach has significant appeal in standard projects, design projects are generally more concerned with progress relative to allowed time than to available budget, and physical measures are not likely to be available. One alternative is to use the estimated duration of the team's activities instead of dollar budgets, together with a simple rule for tracking progress. One simple rule is that a team can immediately claim 33 percent progress for an activity if work on the activity has begun. However, the team gains no additional progress on the task until the activity is completed. The team receives the remaining 67 percent credit for the task only on activity completion. In no case is a team given more than 100 percent credit for a task, no matter how long that activity really takes. Further, the team gets full credit when the work is done, regardless of how long it actually took. (This convention clearly places a premium on careful and accurate decomposition of the work in the WBS.)

Consider Figure 7.8, in which we show a modified PCM for the beverage container design project. Each of the tasks used in the activity network has been included, except for the summary tasks, such as understanding customer requirements and evaluating alternatives. (The detailed breakouts for these have been included instead.) The PCM shows the planned or budgeted duration for each task, the percent of the total project that the task accounts for, and its status. In those cases where a task has either been started or completed, credit toward the overall project is given. In cases where there is no progress, no credit is given. Three observations are in order. First, the project manager or team leader could give a more exact percentage of completion than the 0 percent, 33 percent, 100 percent used in this example, or could do so for selected tasks. Remember that it is the team that chooses the values for the simple, standard rule. Second, the team can compare the progress achieved thus far with the overall time allocated for the project. For example, if the design project were in the fourth week of a ten-week project, the PCM indicates that the project is more or less on plan. If the team was in the eighth week, this PCM is cause for alarm. Third and last, we note that if the team has done a good job of determining the nature and duration of the tasks required to finish the project, this PCM and this method will allow them to monitor their work. If they have not, this method is simply an illusion.

7.8 MANAGING THE XELA-AID CHICKEN COOP PROJECT

As previously noted, one of our illustrative examples is based on a student design project that was part of a one-semester, introductory engineering design course given at Harvey Mudd College. Since only a limited number of project management tools could be successfully applied in this context, the student teams asked to design a chicken coop for the residents of San Martin Chiquito, Guatemala, were asked to develop and submit only WBSs. Two examples are given in Figures 7.9 and 7.10. Several comments are in order.

First, the two teams chose quite different approaches, with one team using a graphical presentation and the other a tabular chart. The WBS in Figure 7.9, aside from the team's oversight in not including lines, is quite similar to that in Figure 7.1. We can see that this WBS has several problems, the most obvious of which is that several of the subtasks are not actually breakdowns of the higher order tasks. Note that "Analyze Function Requirements" has an associated subtask, "Black Box Functional Analysis." Absent another subtask, the

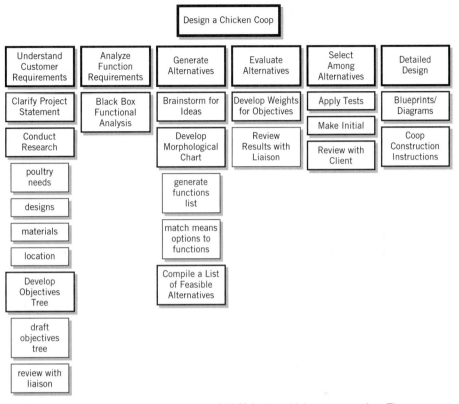

FIGURE 7.9 A work breakdown structure (WBS) for the chicken coop project. There are several errors in this WBS that are commonly made by less experienced project managers. Can you spot them?

one given is simply a restatement of its parent task. Remember that the WBS is a way of breaking down the work into constituent components or smaller jobs.

A second problem with the example in Figure 7.9 is that it is not complete. The team will certainly have some additional management responsibilities, including team meetings and progress reports. The figure doesn't include them. An incomplete WBS will ensure that all other plans that depend on it will also be incomplete.

The task list in Figure 7.10 is more complete, even including the team's estimate of time required to complete each task. It goes so far as to include the time needed to develop the task list (item *iv*) The real problem with this table is that it is a list, not a hierarchy, and so it doesn't show the relationship between tasks. This may seem unimportant to the novice project manager, but understanding the relationships among tasks is an important insight when decisions must be made about the order of activities and, under time pressure, those activities that can be left undone.

	Task	Person Hours
i.	talk to client	5
ii.	read client statement	1
iii.	create objectives tree	6
iv.	create task list	3
v.	revise client statement	7
vi.	oroato function box	4
vii.	talk with client about work above	2
viii.	revise objectives tree, function box, and task list	7
ix.	research Gautemala	7
x.	research chickens and coop designs	16
xi.	other research	4
xii.	create a morphological chart	6
xiii.	combine means on morph chart into coherent designs	4
xiv.	determine the best conceptual combination of means	6
xv.	discuss further alternatives	5
xvi.	discuss with client	2
xvii.	acquire materials for a scale model	8
xviii.	build scale model	12
xix.	build segments of full size replica	20
xx.	documentation of building process	5
xxi.	determine metrics for evaluating design	10
xxii.	evaluation of physical model as well as conceptual design	20
xxiii.	redesign coop if necessary	10
xxiv.	rebuild these segments	10
xxv.	retest coop	5
xxvi.	evaluate coop	5
xxvii.	present design to client	2
xxviii.	make small changes if necessary	5
xxix.	prepare written report	20
xxx.	prepare how-to build instructions	5
xxxi.	prepare oral presentation	20
xxxii.	present to class and client	2
	Total time to complete our project	**245**

FIGURE 7.10 A task list that could serve as the basis for a WBS for the chicken coop project. Note that the team has estimated the time requirements for the project by summing up over all the tasks. If placed in hierarchical order, this could be very easily adapted to a WBS.

7.9 NOTES

Section 7.1: The definition of projects is from (Meredith and Mantel, 1995).

Section 7.2: The underlying conceptual model for the 3S approach is from (Oberlander, 1993). The example in Figure 7.2 is adapted from (Kezsbom, Schilling, and Edward, 1989).

Section 7.3: The text-based WBS example shown in Figure 7.3 is from a sample set included in the software Primavera Project Planner, Release 2.0.

Section 7.4: Linear responsibility charts are explained further in most introductions to project management, for example, (Meredith and Mantel, 1995).

Section 7.5: Activity networks and Gantt charts are covered in greater detail in any introductory text on project management, for example, (Meredith and Mantel, 1995).

Section 7.7: The use of the standard form of the Percent Complete Matrix is given in (Oberlander, 1993). The form as modified draws upon a method given in (CIIP, 1986).

Section 7.8: The WBSs of Figures 7.9 and 7.10 are from (Gutierrez et al., 1997) and (Connor et al., 1997), respectively.

7.10 EXERCISES

7.1 Explain the differences between managing design projects and managing the implementations of design projects. For example, consider the differences between designing a highway interchange and building that interchange.

7.2 Develop a work breakdown structure (WBS) and a linear responsibility chart (LRC) for an on-campus benefit to raise money for the homeless.

7.3 Develop a work breakdown structure (WBS) and a linear responsibility chart (LRC) for an on-campus meeting to be used by the women in the cooperative cited in Exercise 3.5 as they arrive at a nearby airport and travel to your campus. (Hint: You are not allowed to do the courteous thing and pick them up at the airport.)

7.4 Develop a work breakdown structure (WBS) and a linear responsibility chart (LRC) for a project to design a robot that will be entered into a national collegiate competition.

7.5 Develop an activity network for the on-campus benefit of Exercise 7.2.

7.6 Develop an activity network for the robot project of Exercise 7.4.

7.7 Develop a schedule and a budget for the on-campus benefit of Exercise 7.2.

CHAPTER 8

DESIGN FOR . . .

What are some of the consequences of our technical choices?

A **CENTRAL** theme of this book is that engineering design is usually done by teams rather than individuals. This idea reflects the recent experience of engineers in industrial settings throughout the world. Design teams usually include not only engineers, but also manufacturing experts (who may be industrial engineers), marketing and sales professionals, reliability experts, cost accountants, lawyers, and so on. Such teams are concerned with understanding and optimizing the product under development for its *entire life*, including its design, development, manufacturing, marketing, distribution, use, and, eventually, disposal. Concern with all of these areas and their impact on the design process has come to be known as *concurrent engineering*. While it is beyond the scope of many smaller projects (e.g., the student design examples described earlier) to apply concurrent engineering techniques, it is important that engineering designers be aware of concurrent engineering and its implications for their design work.

One of the most important aspects of engineering in the modern commercial setting is the realization that the audience for a good design includes those who will build and maintain the designed artifact. Another important feature is that designs must usually meet economic or cost-related targets. These aspects are part of a more general notion; engineers have always sought to realize various desirable attributes, to some degree, in their designs. This is often referred to as "design for *X*," where *X* is an attribute such as manufacturing, maintainability, reliability, or affordability. (Designers and engineers also refer to them with a different name, the *-ilities*, because many of these desirable attributes are expressed as nouns that have an "-ility" suffix.)

As designers, we can use the idea of the life cycle of a product to guide us through some of the *X*'s. Since most products are designed to be built, sold, used, and then disposed of, we look first at design for manufacturing and assembly, and then at design for affordability, design for reliability and maintainability, and lastly, at design for sustainability. We will see that these and related concepts can be summarized by the idea of *quality*, so we will briefly look at *quality function deployment* (QFD), one approach to design for quality.

8.1   DESIGN FOR MANUFACTURING AND ASSEMBLY: CAN THIS DESIGN BE MADE?

In many cases, a designed artifact will be produced or manufactured in quantity. In recent years, companies have come to learn that the design of a product can have an enormous impact on the cost of producing it, its resulting quality, and its other characteristics. Toward this end, globally competitive industries such as the automotive and consumer electronics industries routinely consider how a product is manufactured during the design process. A significant driver of this concern is the large number of products being manufactured, which allows for economies of scale, which we discuss in Section 8.2. Further, the time it takes to get a product to the consumer, known as the *time to market*, defines a company's ability to shape a market. Design processes that incorporate manufacturing issues can be key elements in speeding products through to commercial production.

8.1.1   Design for manufacturing (DFM)

Design for manufacturing (DFM) is design based on minimizing the costs of production and/or the time to market for a product, while maintaining an appropriate level of quality. The importance of maintaining an appropriate level of quality cannot be overstated because without an assurance of quality, DFM is reduced to simply producing the lowest cost product.

DFM begins almost inevitably with the formation of the design team. In commercial settings, design teams committed to DFM tend to be multi-disciplinary, including engineers, manufacturing managers, logistics specialists, cost accountants, and marketing and sales professionals. Each brings particular interests and experience to a design project, but all must move beyond their primary expertise to focus on the project itself. In many world-class companies, such multidisciplinary teams have become the de facto standard of the modern design organization.

Design for manufacturing is an iterative process with designers and builders.

Manufacturing and design tend to interact iteratively during product development. That is, the design team learns either of a problem in producing a proposed design or of an opportunity to reduce design cost or timing, as a result the team reconsiders its design. Similarly, a design team may be able to suggest alternative production approaches that lead manufacturing specialists to restructure processes. In order to achieve fruitful and synergistic interaction between the manufacturing and design processes, it is important that DFM be considered in each and every one of the design phases, including the early conceptual design stages.

One basic methodology for DFM consists of six steps:

1. estimate the manufacturing costs for a given design alternative;
2. reduce the costs of components;
3. reduce the costs of assembly;
4. reduce the costs of supporting production;
5. consider the effects of DFM on other objectives; and
6. if the results are not acceptable, revise the design once again.

This approach clearly depends upon an understanding of all the objectives of the design; otherwise the iteration called for in Step 6 cannot occur meaningfully. An understanding of the economics of production (some of which are discussed in Section 8.2) is also required. In addition to these, however, there are engineering and process decisions that directly influence the cost of producing a product. Some processes for shaping and forming metal, for example, cost much more than others and are called for only to meet particular engineering needs. Similarly, some types of electronic circuits can be made with high-volume, high-speed production machines, while others require hand assembly. Some design choices that require higher costs for small production runs may actually be less expensive if the design can also be used for another, higher-volume purpose. In each of these instances, a successful design can be completed only by combining deep knowledge of manufacturing techniques with deep design experience.

8.1.2   Design for assembly (DFA)

Design for assembly is a related, but formally different brand of design for *X*. Assembly refers to the way in which the various parts, components, and subsystems are joined, attached, or otherwise grouped together to form the final product. Assembly can be characterized as consisting of a set of processes by which the assembler (1) handles parts or components (i.e., retrieves and positions them appropriately relative to each other), and (2) inserts (or mates or combines) the parts into a finished system or subsystem. For example, assembling a ball point pen might require that the ink cartridge be inserted into the tube that forms the handgrip, and that caps be attached to each end. This assembly process can be done in a number of ways, and the designer needs to consider approaches that will make it possible for the manufacturer to reduce the costs of assembly while maintaining high quality in the finished product. Clearly, assembly is a key aspect of manufacturing and must be considered either as part of design for manufacturing or as a separate, yet strongly related design task.

Because of its central place in manufacturing, a great deal of thought has been put into development of guidelines and techniques for making assembly more effective and efficient. Some of the approaches typically considered are:

1. *Limiting the number of components to the fewest that are essential to the working of the finished product.* Among other things, this implies the designer differentiates between parts that could be eliminated by combining other parts and those that must be distinct as a matter of necessity. The usual issues for this are to identify:

- parts that must move relative to one another;
- parts that must be made of different materials (e.g. for strength or insulation); and
- parts that must be separated for assembly to proceed.

2. *Using standard fasteners and/or integrating fasteners into the product itself.* Using standard fasteners also allows an assembler to develop standard routines for component assembly, including automation. Reducing the number and types of fasteners allows the assembler to construct a product without retrieving as many components and parts. The designer should also consider that fasteners tend to induce stress concentrations and may, thus, cause reliability concerns of the sort we will discuss in Section 8.3.

3. *Designing the product to have a base component on which other components can be located*, and designing for the assembly to proceed with as little motion of the base component as possible. This guideline enables an assembler (whether human or machine) to work to a fixed reference point in the assembly process and minimize the degree to which the assembler must reset reference points.

Consider the number of parts in a design, and how they will be assembled.

4. *Designing the product to have components that facilitate retrieval and assembly.* This may include elements of detailed design that, for example, reduce the tendency of parts and subassemblies to become tangled with one another, or designing parts that are symmetric, so that once retrieved they can be assembled without turning to a preferred end or orientation.

5. *Designing the product and its component parts to maximize accessibility, during both manufacturing and subsequent repair and maintenance.* While it is important that the components be efficient in their use of space, the designer must balance this need with the ability of an assembler or repairer to gain access to and manipulate parts, both for initial fabrication and later replacement.

While these guidelines and heuristics represent only a small set of the design considerations that make up design for assembly, they do provide a starting point for thinking about both DFA and DFM.

8.1.3 The Bill of Materials (BOM)

Effective design for manufacturing also requires a deep understanding of production processes, among the most important of which are ways to plan and control inventories. A common inventory planning technique is Materials Requirements Planning (MRP). It utilizes the assembly drawings (discussed in Chapter 6) to develop a Bill of Materials (BOM) and an assembly chart, called a "gozinto" chart, that shows the order in which the parts on the BOM are put together. The BOM is a list of all of the parts, including the quantities of each required to assemble a designed object. We might think of the BOM as being a recipe that specifies (1) all of the ingredients needed, (2) the precise quantities needed to make a specified lot size, and when paired with the "gozinto" chart, (3) the process for putting the ingredients together.

When a company has determined the size and timing of its production schedule, the BOM is used to determine the size and timing of inventory orders. (Most companies now use *just in time* delivery of parts as they try to avoid carrying large inventories of parts that are paid for but not generating revenues until after they are assembled and shipped.) The importance of the assembly drawings and the BOM in managing the production process cannot be overstated. To be effective, the design team must not only develop accurate methods of reporting their design, but the entire organization must be committed to the discipline that any design changes, or *engineering change orders*, will be reported accurately and thoroughly to *all* the affected parties. In Section 8.2, we will see that the BOM is also useful in estimating some of the costs of producing the designed artifact.

A final point to note is that manufacturing concerns include both logistics and distribution, so that these elements have also become an important part of design for manufacturing. One of the major changes in how business is done today is that companies now work toward forging links between the suppliers of materials, the fabricators, and the channels needed to efficiently distribute the finished product. This set of related activities, often referred to as the *supply chain*, requires a designer to understand elements of the entire product cycle. It is beyond our scope to explore the role of supply chain management in design, except to note that in many industries, successful designers understand not only their own production and manufacturing processes, but also those of their suppliers and their customers. This requirement for integrated understanding of commercial processes will surely increase in the future.

8.2 DESIGN FOR AFFORDABILITY: HOW MUCH DOES THIS DESIGN COST?

To be able to afford something is, according to the dictionary, to be able to bear the cost of something or to pay the price for it. In the design context, whether or not we can afford something is an issue that will be faced by the client (e.g., can I afford to make this product? or perhaps, can I afford not to make it?), the manufacturer (e.g., can I afford to make this at a given price?), and the user (e.g., can I afford to buy this product?). Thus, affordability is really about expressing an important dimension of the object or system being designed in terms that all stakeholders recognize and understand, money. This dimension is typically covered by the field known as *engineering economics*. Engineering and economics have been closely linked for almost as long as the two fields have existed. Indeed, economists recognize that engineers were the developers of a number of important elements of economic theory. For example, what economists call utility theory and price discrimination were both first articulated by the 19th century engineer Jules Dupuit, and location theory was developed by a civil engineer named Arthur M. Wellington. In fact, Wellington is credited with the definition of engineering as "the art of doing that well with one dollar which any bungler can do with two." This linking of engineering and economics should come as no surprise to the designer, since it is a rare project for which money really is "no object."

> Virtually all engineering decisions have economic elements.

Engineering economics is concerned with understanding the economic or financial implications of engineering decisions, including choosing among alternatives (e.g., cost-benefit analysis), deciding if or when to replace machines or other systems (replacement analysis), and predicting the full costs of devices over the period of time that they will be owned and used (life cycle analysis). These topics can easily fill entire courses in an engineering curriculum and are well beyond our current scope. However, there are some topics sufficiently important to a designer that they must be briefly introduced, perhaps the most important of which is the *time value of money*. A second important topic is *cost estimation*. Without at least a rudimentary knowledge of these, design and engineering teams are likely to make good choices about designs only by luck.

8.2.1 The time value of money

If someone were to offer us $100 today or $100 one year from now, we would almost certainly prefer to take the money *now*. Having the money sooner offers a number of advantages, including providing us with the ability to invest or otherwise use the money during the intervening year; eliminating the risk that the money might not be available next year;

A dollar today is worth more than the promise of a dollar tomorrow.

and eliminating the risk that inflation might reduce the purchasing power of the money during the next year. This simple example highlights one of the most important concepts in engineering economics, namely, the *time value of money*: Money obtained sooner is more valuable than money obtained later, and money spent sooner is more costly than money spent later.

As indicated, the time value of money captures the effects of both foregone opportunities, referred to as *opportunity costs*, and *risk*. An opportunity cost is a measure of how much the deferred money could have earned in the intervening time. The risk captures both the risk that the money will be worth less (because of inflation) and the risk that the money will simply have become unavailable during the intervening time. Economists and financial professionals bundle together the extent of these risks and their associated lost opportunities in the *discount rate*. A discount rate acts much like an interest rate on a savings account or credit card, except that it views the money as being worth less to us tomorrow because of the risks and missed opportunities. The interest rate on a savings account measures how much a bank is willing to pay for the privilege of using our money in the coming year. An interest rate on a credit card measures how much we must pay the card issuer for the privilege of using their money; these rates vary, with higher charges assigned to customers with poorer credit ratings or limited credit histories. Interest calculations thus typically show dollar amounts increasing from a given time onward or forward. Discount rates typically work in reverse, showing the value today of money that will be available at some point in the future.

Measuring risks and opportunity costs can be a complex process, but as engineers we should remember that design decisions and choices made today will translate into streams of prospective "financial events" that will occur at different times in the future. Some of these financial events are *costs* we will incur (e.g., for manufacturing or distribution), and some are *benefits* (e.g., revenues from sales) we will derive. The more immediate the costs and benefits, the more impact they have on decisions made by clients, users, and designers.

How do we distinguish between "$100 today" and "$100 a year from now" in a rational and consistent way? The answer is that, given a discount rate and a set of future financial events or cash flows, both toward us and away from us, we translate all of our events into a common time frame, either a current time or a future one. Consider again our choice between getting $100 today or $100 next year. If the annual discount rate is 10%, then we would expect to need $110 dollars in a year in order compensate us for not having $100 today (or to purchase then what we could buy today with $100). Then we would be getting the same amount of money if we accept $100 now or $110 a year from now.

We can also work this the other way, asking how much we need today to have the equivalent of $100 a year from now. Next year's $100 is worth the same amount as some amount of money, X, plus the 10% that X could earn in the coming year. That is, $1.10 \times \$X = \$100$. Solving for X shows that taking the $100 in a year is equivalent to getting and investing about $91 today. We won't go through all of the various formulas and arithmetic associated with the time value of money, except to point out that we can carry out discounting calculations as far into the future as we want. The principle is the same. Thus, $100 promised two years from now would be worth even less than $100 were today or $100 promised for next year, since it wouldn't have accounted for being able to use the $100 for both years, or our being able to use the $10 earned in the first year.

Economists have developed standard approaches to discounting money and to determining the present value of future dollars, whether costs or benefits, and vice versa. Application of these formulas can become quite involved when inflation or unusual timing issues are involved, but virtually all such analysis is based on the relationship

$$PV = FV \left(\frac{1}{1+r} \right)^t$$

(8.1)

where PV is the present value of the costs or benefits, FV is the future value, r is the discount rate, and t is the time period over which a cost is incurred or a benefit is realized. Consider again, the decision concerning the worth of $100 next year. In this case, the future value, FV, is $100, the time period, t, is 1 year, and the discount rate, r, is 10% per year, or 0.10. If substitute these values into eqn. (8.1), the present value, PV, will be the $91 found above. In other words, an offer of $100 a year from now would be equivalent to an offer of $9 less today. The ability to translate future costs into present equivalent values, known as *discounting*, is very important in some design projects and may affect how we choose among designs. We turn to that topic now.

8.2.2 The time value of money affects design choices

Imagine for a moment that we are asked to choose between two alternative vehicle designs for a transit agency. Design alternative A has a significantly higher initial purchase price, but design alternative B has higher operating costs over the life of the vehicle. In this case we have to reconcile the different costs at different points in the lifetimes of the two choices. One obvious question is, How big are those cost differences and when do they occur? If design B can be bought for much less money than design A, its higher operating costs may not be as important because they are incurred so much later than the immediate initial saving. If, on the other hand, the operating costs of B are both much higher and occur relatively early in the vehicle's life, then the immediate savings gained by buying design B may be illusory. Making a rational choice in such a case requires that we understand how to properly analyze the time value of money (see Section 8.1.1) because we have to compare all of the costs in equivalent dollar values.

Designs are purchased today **and** used over their lifetimes.

Now, imagine further that our designs have not only different purchase prices and different operating and maintenance costs, but also different expected lifetimes. This means that we have to adjust all of the costs in a way that makes the values equivalent over the same time frame. Engineering economists have developed a methodology for doing that, although we will not provide a detailed description here. Called *equivalent uniform annual costs* (EUAC), it essentially treats all of the alternatives as though they are replaced with a one-for-one swap whenever they wear out. EUAC then transforms the resulting infinite series of replacements into a series of annual payments. The point to be drawn here is simply that the series of future costs and benefits of all design alternatives must be considered over the lifetime of each of the alternatives and then translated into a format that allows us to fairly compare those alternatives. The essential wisdom is that it is insufficient to look only at the initial purchase costs of design alternatives as a way of finding out what designs really cost. A true cost analysis requires consideration of the entire life cycle of a design.

8.2.3 Estimating costs

We took it for granted in the previous section that the cost of the final design is known, as are the operating and maintenance costs over the life of the device. In practice cost estimation is not usually so simple. It requires skills and experience, and it can easily consume an entire text. However, several points are relevant during conceptual design.

It is easy to say that the costs of a design typically include labor, materials, overhead, and profits for various stakeholders. However, this simple statement masks the complexity of detailing or structuring the cost of all but the simplest of artifacts. In many cases, estimating the production and distribution costs of a design is extremely difficult. Here we limit ourselves to describing the principal elements that make up the cost categories listed above.

Labor costs include payments to the employees who construct the artifact, as well as to support personnel who perform necessary but often invisible tasks, such as answering the phone, filling orders, packing and shipping the product, etc. Labor costs also include a variety of *indirect costs* that are less evident because they are not payments made directly to employees. Such indirect costs are called *fringe benefits* because they are typically payments made to third parties on behalf of the employees. The fringe benefits include health and life insurance, retirement benefits, employers' contributions to Social Security, and other mandated payroll taxes. These indirect costs of labor are often neglected or overlooked by designers estimating the cost of a design, yet for many companies they are as much as 50% of direct labor payments or wages.

In Section 8.1 we discussed the importance of the Bill of Materials (BOM) in controlling inventories and managing the manufacturing of items. The BOM is also useful for estimating the material costs associated with various designs. *Materials* include those directly used in building the device, as well as intermediate materials and inventories used in ways that may not be obvious. For example, some inventory is wasted during manufacturing, while other inventory may be classified as being part of work-in-progress. The BOM provides a guideline for the number and type of parts that make up the device or object. The BOM is particularly useful since it is developed directly from the assembly drawings, and so it reflects the designer's final intent.

We must to exercise care when we use a BOM to estimate costs since both labor and materials are subject to *economies of scale*, the idea that the *unit cost* or production cost per (single) item can often be reduced by making many identical copies, rather than just making a few "originals." The genius of Henry Ford's assembly line, wherein he came up with a way to make millions of copies of his cars, is a reflection of the economies of scale. Ford lowered his unit costs and sold all of the cars that he did because many more people could afford to buy them. Of course, Ford realized such economies of scale by developing new technologies, but that's still another story about how engineering and economics interact.

Costs incurred by a manufacturer that cannot be directly assigned to a single product are termed *overhead*. If, for example, a device is made in a factory that also produces twenty other products, the cost of the building, the machines, the janitorial staff, the electricity, etc., must somehow be shared or distributed among all of the 21 items. If each product was priced to ignore these overhead costs, the company would soon find itself unable to pay either for the building or for the services necessary to maintain it. Other elements of overhead include the salaries of executives, who presumably use some of their time to supervise each of the company's activities, and personnel and related costs of needed business functions such as accounting, billing, and advertising. While there are accounting standards that define cost categories and their attributes, precise estimates of overhead costs vary greatly with the structure and practices of the company in question. One company may have only a small number of products and a very lean organization, with most of its costs directly attributed to the products made and sold, and only a small percentage allocated to overhead. In other organizations the overhead can be equal to or greater than the labor costs directly assignable to one or more products. In many universities and colleges, for example, the overhead rate associated with research (which pays for laboratories, support staff, college presidents, and other essentials) runs as high as 65 percent of the researchers' salaries and benefits. The key point is that estimating the costs of producing a design requires careful consultation with clients or their suppliers.

Cost estimates produced during the conceptual stage of a design project are often quite inaccurate when compared with those made for detailed designs. In heavy construction projects, for example, an accuracy of ±35 percent is considered acceptable for initial estimates. However, this tolerance of inaccuracy should not be taken as a license to be sloppy or casual in early cost estimating.

In practice, each of the engineering disciplines has its unique approaches to cost estimating, and these approaches will often be captured by some useful heuristics or *rules of thumb*, or general guidelines, that are most relevant at the conceptual design stage. In civil engineering, for example, the *R. S. Means Cost Guide* provides cost estimates per square foot for the various elements in different kinds of construction projects. The *Richardson's Manual* offers similar information for chemical plant and petroleum refinery projects. On the other hand, costs per square inch may be more relevant for printed circuit board designs. In each of the disciplines, we should carefully consult with experienced professionals to estimate costs successfully, even at the more general levels we need to make conceptual design choices.

Finally on cost estimation, we want to highlight the distinction between the cost of designing an artifact, on the one hand, and the cost of manufacturing and distributing it, on the other. In many cases, the cost of design is a relatively small part of the final project cost, such as in the case of a dam or other large structure. Notwithstanding that, however, most clients expect that a design team will correctly estimate its own costs and budget them accurately. Thus, even when costing out the design activity, an effective design team will seek to understand and control its costs accurately.

8.2.4 Costing and pricing

While costs affect profitability, prices are based on value, not costs.

Finally, it should be noted that while costing is an important element in the *profitability* of a design, it is generally *not* a key factor in the *pricing* of the artifact. This seeming contradiction is easily explained by noting that gross profits (i.e., profits before taxes and other considerations) are simply the net of revenues minus costs. As such, costs are an important element in the profit equation. Revenues, on the other hand, are determined by the price multiplied by the number of items sold. For most profit-maximizing firms, prices are not set on the basis of costs, but rather in terms of what the market is willing to pay. Some examples will illustrate this.

Consider a high-quality graphite tennis racket. It may command a price of hundreds of dollars when initially introduced. It is clear upon inspection, however, that the costs of making such a racket are nowhere near such figures. The materials for the racket may cost just a few dollars, the labor is virtually negligible, and the costs of technological developments are relatively modest when *amortized* (spread out) over the many thousands of sales of the racket. The distribution costs are clearly no different for high-end rackets than for the $10 rackets found at discount stores. However, since there is an obvious demand for such expensive rackets, their prices are set high because there are customers are willing to pay high prices. Indeed, the role of marketing professionals on the design team usually includes identifying design attributes that will cause consumers to pay a premium for a designed product.

This example also serves to highlight an aspect of *reliability*. A manufacturer can offer a virtually lifetime replacement guarantee if the costs of manufacture are so far below the selling the price. Thus, the high-quality service of certain brands reflects the disparity between their price and their cost structure.

A similar example can be found in the airline industry. Here the provider of the service faces essentially the same costs regardless of whether a flight is almost full or almost empty. This explains why airlines are willing to offer certain deeply discounted fares at some times, and almost no discounts at others (such as holidays). It also explains why the airlines are willing to invest heavily in modeling and tracking the wide variety of fare options they make available.

In some industries a convention has arisen to compensate designers or providers of certain products on a "cost-plus" basis. For example, most large public works projects, such as highways or dams, are built on the basis of the costs of the contractor or designer plus an additional percentage as a profit allowance. While this is common practice in some cases, the norm in the private sector is to select prices to maximize profits, not to simply tack on a profit factor.

The point is that an engineering designer must control the costs of the design to ensure that the objectives of the client and users are realized. Beyond that, however, the ultimate profitability of an artifact may turn out to be beyond a designer's control, as in the cases of pet rocks and beanie babies.

8.3 DESIGN FOR RELIABILITY: HOW LONG WILL THIS DESIGN WORK?

Most of us have a personal, visceral understanding of reliability and unreliability as a consequence of our own experience with everyday objects. We say that the family car is unreliable, or that a good friend is very reliable—someone we can count on. While such informal assessments are acceptable in our personal lives, we need greater understanding and accuracy when we are functioning as engineering designers. Thus, we now describe how engineers approach reliability, along with its sister concept, maintainability.

8.3.1 Reliability

To an engineer, reliability can be defined as "the probability that an item will perform its function under stated conditions of use and maintenance for a stated measure of a variate (time, distance, etc.)." This definition has a number of elements that warrant further comment. The first is that we can properly measure the reliability of a component or system *only*

Reliability is the probability that an item will function under stated conditions for a stated measure of usage

under the assumption that it has been or will be used under some particular set of usage and maintenance conditions. The second point is that the appropriate measure of use of the design, called the *variate*, may be something other than time. For example, the variate for a vehicle would be miles, while for a piece of vibrating machinery the variate would be the number of cycles of operation. Third, we must examine reliability in the context of the functions discussed in Chapter 4, which emphasizes the care we should take in developing and defining the functions that a design must perform. Finally, note that reliability is treated as a probability, and hence can be characterized by a distribution. In mathematical terms, this means that we can express our expectations of how reliable, safe, or successful we expect a product or a system to be in terms of a cumulative distribution function or a probability density function.

In practice, our use of a probabilistic definition enables us to consider reliability in the context of the opposite of success, that is, in terms of *failure*. In other words, we can frame our consideration of reliability in terms of the probability that a unit will fail to perform its functions under stated conditions within a specified window of time. This requires us to consider carefully what we mean by failure. *British Standard 4778* defines a failure as "the termination of the ability of an item to perform a required function." This definition, while helpful at some level, does not capture some important subtleties that designers must keep in mind: It doesn't capture the many kinds of failures that can afflict a complex device or system, their degree of severity, timing, or effect on the performance of the overall system.

For example, we find it useful to distinguish between *when* a system fails and *how* it fails. If the item fails when in use, the failure can be characterized as an *in-service failure*. If the item fails, but the consequences are not detectable until some other activity takes place, we refer to that as an *incidental failure*. A *catastrophic failure* occurs when a failure of some function is such that the entire system in which the item is embedded fails. For example, if our car breaks down while we're on a trip and needs a repair in order for us to complete the trip, we would call that an in-service failure. An incidental failure might be some part that our favorite mechanic suggests we replace during the routine servicing of our car. A catastrophic, accident-causing failure might follow from the failure of a critical part of the car while we are driving at freeway speeds. Each type of failure has its own consequences for the users of the designed artifact, and so must be considered carefully by designers.

We often specify reliability in part by using measures such as the mean time between failures (MTBF), or miles per in-service failure, or some other variate or metric. However, we should note that framing the definition of reliability in terms of probabilities gives some insight into the limitations inherent in such measures. Consider the two failure distributions shown in Figure 8.1. These two reliability probability distributions have the same mean (or average), that is, $\text{MTBF}_a = \text{MTBF}_b$, but they have very different degrees of dispersion (typically measured as the variance or standard deviation) about that mean. If we are not concerned with both mean and variance, we may wind up choosing a design alternative that is seemingly better in terms of MTBF, but much worse in terms of variance. That is, we may even choose a design for which the MTBF is acceptable, but for which the number of early failures is unacceptably high.

One of the most important reliability issues for a designer is how the various parts of the design come together and what the impact is likely to be if any one part does fail. Consider, for example, the conceptual sketch of the *series system* design, shown in Figure 8.2.

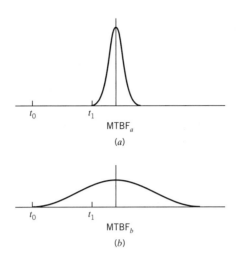

FIGURE 8.1 Failure distributions (also called probability density functions) for two different components. Note that both curves have the same value of MTBF, but the dispersions of possible failures differ markedly. The second design (sketch b) would be viewed as less reliable because more failures would occur during the early life of the component (i.e., during the time interval $t_0 \leq t \leq t_1$).

It is a chain of parts or elements, the failure of any one of which would break the chain, which in turn will cause the system to fail. Just as a chain is no stronger than its weakest link, a series system is no more reliable than its most unreliable part. In fact, the reliability—or probability that the system will function as designed—of a series system whose individual parts have reliability (or probability of successful performance) is given by:

$$R_S(t) = R_1(t) \cdot R_2(t) \cdot \ldots \cdot R_n(t)$$

or

$$R_S(t) = \prod_{i=1}^{n} R_i(t)$$

(8.2)

where is the reliability of the entire series system, and $\prod_{i=1}^{n}$ is the product function. We see from eqn. (8.2) that the overall reliability of a series system is equal to the product of all of the individual reliabilities of the elements or parts within the system. This means that if any one component has low reliability, such as the proverbial weak link, the entire system will have low reliability and the chain will break.

FIGURE 8.2 This is a simple example of a *series system*. Each of the elements in the system has a given reliability. The reliability of the system as a whole can be no higher than that of any one of the parts because the failure of any one part will cause the system to cease operating. What is the reliability of this system as calculated with Eqn. (8.2)?

Designers have long understood that redundancy is important for dealing with the weakest-link phenomenon. A *redundant system* is one in which some or all of the parts have backups or replacement parts that can substitute for them in the event of failure. Consider the conceptual sketch of the *parallel system* of three parts or elements shown in Figure 8.3. In this simple case, each of the components must fail in order for the system to fail. The reliability of this entire parallel system is given by:

$$R_P(t) = \left[\left(1 - R_1(t)\right) \bullet \left(1 - R_2(t)\right) \bullet \ldots \bullet \left(1 - R_n(t)\right)\right]$$

or

$$R_P(t) = 1 - \prod_{i=1}^{n}\left[1 - R_i(t)\right]$$

(8.3)

We see from eqn. (8.3) that the reliability of this parallel system (i.e., the probability that the parallel system will operate successfully) is now such that if any one of the elements functions, the system will still function.

Redundancy usually increases both reliability and costs.

Parallel systems have obvious advantages in terms of reliability, since all of the redundant or duplicate parts must fail in order for the system to fail. Parallel systems are also more expensive, since many duplicate parts or elements are included only for contingent use, that is, they are used only if another part fails. For this reason, we must carefully weigh the consequences of failure of a part against failure of the system, along with costs attendant to reducing the likelihood of a failure. In most cases designers will opt for some level of redundancy, while allowing other components to stand alone. For example, a car usually has two headlights, in part so that if one fails the car can continue to operate safely at night. The same car will usually have one radio, since its failure is unlikely to be catastrophic. The mathematics of combining series and parallel systems is beyond our scope, but we clearly have to learn and use them to design systems that have any impact on the safety of users.

Designers can consider modes of failure and develop estimates of reliability only if they truly know how components might fail. Such knowledge is gained by performing experiments, analyzing the statistics of prior failures, and by carefully modeling the underlying physical phenomena. Designers lacking deep experience in understanding component failure should consult experienced engineers and designers, users, and the client in order to

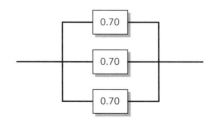

FIGURE 8.3   This is a simple example of a *parallel system*. Note that every one of the components must fail in order for the system to cease working. While such a system has high reliability, it is also expensive. Most designers seek to incorporate such redundancy when necessary, but look for other solutions wherever possible. How does the reliability of this parallel system, as calculated with Eqn. (8.3), compare with that of the series system in Figure 8.1?

ascertain that an appropriate level of reliability is being designed. Often the experiences of others allows a designer to answer reliability questions without performing a full set of experiments. For example, the appropriateness of different kinds of materials for various designs can be discussed with materials engineers, while properties such as tensile strength and fatigue life are documented in the engineering literature.

8.3.2 Maintainability

Our understanding of reliability also leads us to conclude that many of the systems that we design fail if they are used without being maintained, and that they may need some amount of repair even when they are properly maintained. This fact of life leads engineers to consider how to best design things so that the necessary maintenance can be performed effectively and efficiently. Maintainability can be defined as "the probability that a failed component or system will be restored or repaired to a specific condition within a period of time when maintenance is performed within prescribed procedures." As with our definition of reliability, we can learn from this definition.

First, maintainability depends upon a prior specification of the condition of the part or device and on any maintenance or repair actions, which are part of the designer's responsibilities. Second, maintainability is concerned with the time needed to return a failed unit to service. As such, the same concerns regarding inappropriate use of a single measure, (e.g., the mean time to repair (MTTR)), come into play.

Designing for maintainability requires the designer to take an active role in setting goals for maintenance, such as times to repair, and in determining the specifications for maintenance and repair activities in order to realize these goals. This can take a number of forms, including:

- selecting parts that are easily accessed and repaired;
- providing redundancy so that systems can be operated while maintenance continues;
- specifying preventive or predictive maintenance procedures; and
- indicating the number and type of spare parts that should be held in inventories in order to reduce downtime when systems fail.

There are costs and consequences in each of these design choices. For example, a system designed to have increased redundancy in order to limit downtime while maintenance is ongoing, as is the case for an air traffic control system, will have very large attendant capital costs. Similarly, the cost of carrying inventories of spare parts can be quite high, especially if failures are rare. One strategy that has been increasingly adopted in many industries is to work toward making parts standard and components modular. Then spare parts inventories can be used more flexibly and efficiently, and components or subassemblies can be easily accessed and replaced. Any removed subassemblies can be repaired while the repaired system has been returned to service.

If high maintainability has been established as a significant design objective, design teams must take active steps in the design process to meet that goal. As such, a design team should ask itself what maintenance actions (e.g., service) reduce failures, what elements of the design support early detection of problems or failures (e.g., inspection), and what elements ease the return of failed items to use (e.g., repair). While no one would intentionally design systems to make maintenance more difficult, the world is fraught with examples in which it is difficult to believe otherwise, including a new car in which the owner would have to remove the dashboard just to change a fuse!

8.4 DESIGN FOR SUSTAINABILITY: WHAT ABOUT THE ENVIRONMENT?

Many people have come to hold negative views of technology and engineered systems because of the realization that one generation's progress may produce an environmental nightmare for the next. There are certainly enough examples of short-sighted projects (such as irrigation systems that created deserts or flood control schemes that eliminated rivers entirely) that responsible engineers can feel at least some anxiety about what *their* best ideas might eventually produce. The engineering profession has come to appreciate these concerns over the past several decades, and incorporated environmental responsibility directly into the ethical obligations of engineers. The American Society of Civil Engineers, for example, specifically directs engineers to "strive to comply with the principles of sustainable development"; the American Society of Mechanical Engineers Code of Ethics includes Canon 8: "Engineers shall consider environmental impact in the performance of their duties." A number of tools to understand the environmental effects are being introduced into engineering design in order to help with these issues and obligations. Some even take on the force of law, such as the need for environmental impact statements for certain projects. Environmental life cycle assessments (LCAs) are increasingly used, so we note some of the key environmental challenges facing designers and describe LCAs.

> Engineers have an ethical obligation to consider the environmental consequences of the things they design.

8.4.1 Environmental issues and design

Environmental concerns relevant to design can be organized in any number of ways. Transportation engineering texts, for example, concern themselves with the impacts of engineered systems on water and air quality. Alternatively, electrical engineering texts consider the effects of power generation and transmission or focus on the particular environmental effects of some of the solvents and other chemicals associated producing chips or printed circuit boards. A more general approach is to think in terms of particular aspects of the environment and then consider the likely short- and long-term consequences of design alternatives.

We can often characterize the environmental implications of a design in terms of the effects on air quality, water quality, energy consumption, and waste generation. In each case, we need to address both short-term issues, which will likely come up as part of the economic effects, and long-term issues, which may not. Unfortunately, experience shows that the long-term effects of our design choices often completely overwhelm short-term benefits.

Air quality immediately springs to mind when listing environmental concerns related to design. Urban areas have tremendous smog problems, small towns often have an industry with a large smokestack, and national forests are experiencing loss of habitat due to acid rain and other air quality issues. It is important to realize that these enormous problems often begin with relatively small emissions from various steps in the production of everyday objects. Each mile we drive in a standard internal combustion engine powered car adds a tiny bit of particulate matter, nitrous oxide, and carbon monoxide to the atmosphere. In addition, refining the fuel, smelting the steel, and curing the rubber for the tires add further emissions to the air. Less obvious but similar air quality problems result from the production of everyday materials in paper bags and plastic toys. In other words, designers concerned about the environment must consider both the manufacture of the product and its use.

Environmentally conscious engineers should also concern themselves with issues of *water quality* and *water consumption*. We take the availability of clean water for granted. In fact, many of the world's major bodies of water are already under stress from overuse and pollution. As with air quality, this is a direct result of the multiple uses made of our water supplies. Many states have experienced severe droughts in the recent years, and in the southwestern United States, water is becoming the single biggest environmental constraint on further growth. Effective designers must consider and calculate the water requirements for producing and using their designs. Estimating changes to water resulting from particular designs is of great significance. These can include changes in water temperature (which for large processes can affect fish and other parts of the ecosystem) and the addition of chemicals, particularly hazardous or long-lived compounds.

The production and use of designed systems needs *energy*. However, the energy demands of a system can be much higher than designers realize, or may come from sources that are particularly problematic environmentally. Several years ago, California faced an energy crisis that led to sporadic blackouts. Design choices about common household appliances such as refrigerators affect an increasingly energy-starved world. The variety of sizes, shapes, and levels of efficiency of refrigerators highlights the many design choices made by engineers and product design teams. Beneath the surface of such devices, however, there are further design choices made by engineers while generating and selecting alternatives. The principal energy consumer in a refrigerator is the compressor, which can be made more energy-efficient by judicious selection of components. Within the refrigerator walls, the use of insulation materials has a tremendous effect on how well cold temperatures are retained. Even door designs and their placement affects how much energy a refrigerator consumes. Designers must approach such projects systematically, applying all of the skills and techniques learned in their engineering science courses, and accounting for the consequences of their design choices.

Products must be disposed of after fulfilling their useful life. In some cases, perfectly good designs become serious disposal problems. For example, consider the wooden railroad tie used to secure and stabilize trains tracks and distribute the loads into the underlying ballast. Properly maintained and supported, ties treated with creosote typically last more than 30 years, even under heavy loads and demanding weather conditions. Not surprisingly, most railroads use such ties. At the end of their lives, however, the same chemical treatment that made them last so long creates a major disposal problem. Improperly disposed of, the chemicals can leach into water supplies and be harmful to living things. The ties also emit highly noxious, even toxic fumes when burned. Thus, managing the *waste streams* associated with products and systems has become an important consideration in contemporary design. A great solution to one problem has become a problem in itself. The railroad industry has sponsored a number of research projects to explore ways to reuse, recycle, or at least better dispose of used ties, but the results remain to be seen.

Sometimes the market fails to support the planned post-consumer disposal, even for products designed to be recyclable or reusable. Recycling is certainly the intended use of many paper and plastic products, for example, but many cities find it difficult to successfully dispose of recycled paper and so are forced to place it in landfills. Battery companies tried to develop recycling facilities to capture and control metals and other dangerous waste products, but the small and omnipresent nature of batteries mades this difficult.

8.4.2 Environmental life cycle assessment

Life cycle assessment (LCA) was developed to help understand, analyze and document the full range of environmental effects of design, manufacture, transport, sale, use and disposal of products. Depending on the nature of the LCA and the product, such analysis begins with the acquisition and processing of raw materials (such as petroleum drilling and refining for plastic products or foresting and processing of railroad ties), and continues until the product has been reused, recycled, or placed in a landfill. LCA has three essential steps:

- *Inventory analysis* lists all inputs (raw materials and energy) and outputs (products, wastes, and energy), as well as any intermediate outputs.
- *Impact analysis* lists all of the effects on the environment of each item identified in the inventory analysis, and quantifying or qualitatively describing the consequences (e.g., adverse health effects, impacts on ecosystems, or resource depletion).
- *Improvement analysis* lists, measures, and evaluates the needs and opportunities to address adverse effects found in the first two steps.

Obviously, one of the keys in LCA is the setting of assessment boundaries. Another is the determination of appropriate measures and data sources for conducting the LCA. Designers cannot expect to find good, consistent data for all elements in the LCA, and so they must reconcile information from multiple sources. Because of differing boundaries, data sources, and reconciliation techniques, analysts may produce different figures for the overall effects of a product, even when acting in good faith. To address this issue, it is particularly important to list all assumptions made and to document all data sources used.

Currently, LCA is still in its earliest stages of development as a tool for engineering designers (and others concerned with the environmental effects of technologies). Notwithstanding its youth, however, LCA is already a useful conceptual model for design, and is likely to become increasingly important for the evaluation of engineered systems.

8.5 DESIGN FOR QUALITY: BUILDING A HOUSE

Quality unites almost all of the elements of conceptual design.

In a certain sense, all the previous *X*'s we have looked at can be considered dimensions of *design for quality*. Quality itself has been defined in a variety of ways, some very brief, some very complicated. One of our favorites is also one of the simplest: *Quality* is "fitness for use," that is, it is a measure of how well a product or service meets its required or desired specifications. By this definition, much of the problem definition activities discussed in Chapters 2, 3, and 4 are aimed at determining what a "quality" design requires. A design will generally considered a high-quality design if it satisfies all constraints, is fully functional within the desired performance specifications, and meets the objectives as well or better than alternative designs. In that sense, all of the work we have done in conceptual design is directed to design for quality.

Having said that, however, it is often difficult to tie together all of the elements of a good design. This difficulty can come from problems at the design level, which is our primary focus, or at the implementation level, as in manufacturing or distributing seemingly good products. Designers and manufacturers have developed a variety of tools and techniques to address this difficulty and improve the quality of their products. These include process improvement techniques such as *flowcharting* and *statistical process control* (SPC), external comparisons such as *benchmarking* against other high quality products, and improving the distribution and delivery of finished goods, known as *supply chain management*.

One of the most important tools used by many designers is *quality function deployment* (QFD). QFD uses—and is also referred to—as the *house of quality*, a matrix that combines information about stakeholders, desirable characteristics of designed products, current designs, performance metrics, and tradeoffs. Figure 8.4 shows the general structure of a house of quality and illustrates how the house metaphor developed. The *Who* in the figure refers to the stakeholders in the design process, that is, client(s), users, and other affected parties. The *Whats* correspond to the desirable design attributes objectives, and in some cases, to functions. The *Now* in the house are existing products or designs. They are typically found as a result of research that has been conducted during the problem definition phase, and they are used for benchmarking proposed designs. The *Hows* in the house of quality refer to the metrics and specifications used to measure how well an objective or function has been met. In some versions of the house of quality, some of the functions are placed in the *Hows* part of the matrix, particularly if qualitative measures are being used. *How muches* refers to the goals or targets for the *Whats*. In each of the remaining sections, the relationships, values, or trade-offs among these elements are displayed. For example, relationships among ways to realize the desired attributes are exposed in the roof of the house. For example, a device made more reliable with redundant systems will also be more costly, so we might put a minus sign in the box corresponding to these measures.

Consider the example in Figure 8.5. This simple house of quality is used to explore a housing for a laptop computer. Many people have recently begun to use laptops both while

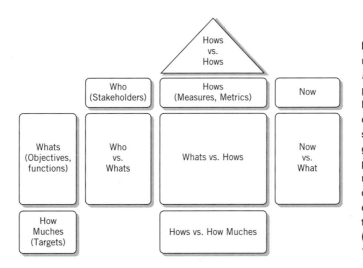

FIGURE 8.4 An elementary abstraction of a *house of quality* that displays and relates stakeholder interests, desired design attributes, measures and metrics, targets, and current products. The "house" reports values for these quantities and helps designers explore relationships among them. (Adapted from Ullman, 1997.)

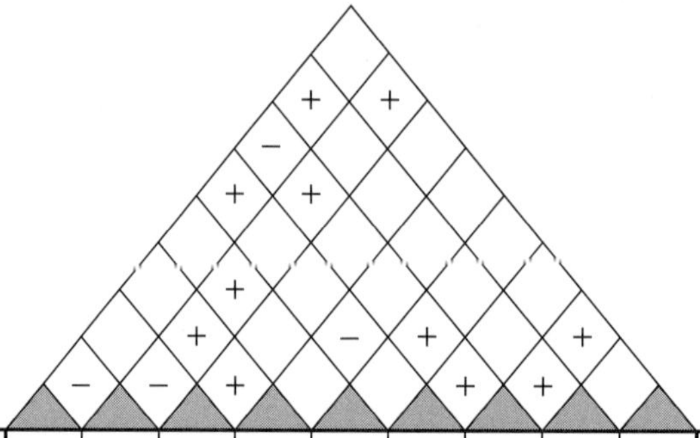

	Traveling Users	Office Users	Manufacturing	Weight (oz)	Cost of Materials ($)	Cost of Assembly ($)	Number of Parts (#)	Conforms to IEEE Std (Y/N)	Yield Strength (MPa)	Rank in Focus Group (#)	Number of Ports or Cards (#)	Postconsumer Reusable Percent (%)	Current Laptop Casing	Current Desktop Casing
Lightweight	H	L	L	●									1	2
Inexpensive	M	H	H		●	●	○						2	1
Attractive	H	M	L							●	○	□	1	2
Durable	H	M	L	○			□		●				2	1
Adaptable	L	M	H								●		2	1
Safe	H	H	L					●					2	1
Recyclable	M	M	M									●	2	1
			Target	16	20	10	2	Y	50	1	4	50		

Priorities:
  H: High
  M: Medium
  L: Low

Relationship:
  ● Strongly Related
  ○ Moderately Related
  □ May Be Related/Weakly Related

FIGURE 8.5 A first-draft house of quality for the design of the housing of a computer that would be used as a laptop and in the office. Note that different users may have different priorities, and the roof of the house helps identify trade-offs between various measures and attributes.

traveling and in the office. A computer maker might want to explore the design space for computer housings that satisfy both office and laptop computers. The stakeholders include traveling users, office users, and the manufacturer's production group. In the *Who vs. Whats* section of Figure 8.4, we see that travelers place a high priority on physical characteristics, such as lightweight and durable, while office users are more concerned with cost and adaptability. We might imagine two existing designs, one a standard laptop case and the other a standard desktop/tower casing. The *Whats vs. Hows* section shows the relationship between the various metrics and the attributes of a "good" design. Notice, for example, that cost of raw materials and cost of assembly are both strongly related to inexpensive, while number

of parts is only modestly so. Similarly, the number of cards and ports that can be accepted is also modestly related to inexpensive, since these require additional assembly work or more parts. *Now vs. What* is the result of benchmarking the two existing design choices, and highlights the possibility that a "universal" housing might be able to satisfy more users in total if it can address the shortcomings of either design. Finally, the roof of the house shows some of the relationships and tradeoffs that designers will need to consider. Making the case lighter, for example, is likely to trade off negatively with resistance to forces. Increasing the number of parts is likely to result in higher costs to assemble.

This simple example shows that the house of quality can help tie together many of the concepts we have considered throughout this book. An important question is: When should QFD be introduced into a design process? Virtually all proponents of the house of quality acknowledge that it requires a good deal of time and effort. Our own experience suggests that it is a useful way to plan to gather information, to organize that information as it becomes available, and to foster and enhance discussion with the design team and with stakeholders. However, as with some of the other tools, it should not be taken as an algorithm that produces a hard-and-fast decision, as it cannot produce better results than the materials on which it works!

8.6 NOTES

Section 8.1: This section draws heavily upon (Pahl and Beitz, 1996) and (Ulrich and Eppinger, 1995). In particular, the six-step process is a direct extension of a five-step approach in (Ulrich and Eppinger, 1995), with the iteration made explicit. Our discussion of DFA is adapted from (Dixon and Poli, 1995) and (Ullman, 1997); rules of assembly are cited in many places but are generally derived from (Boothroyd and Dewhurst, 1989). The analogy of the BOM to a recipe is taken from (Schroeder, 1993).

Section 8.2: There are any number of excellent texts in engineering economics that can be used to go further with these topics. The cost estimating material draws upon classnotes prepared by our colleague Donald Remer for his cost estimating course and on (Oberlander, 1993). A view of engineering costing based more on accounting may be found in (Riggs, 1994). The relationship between pricing and costs is discussed in (Nagle, 1987) and (Phlips, 1985).

Section 8.3: The definition of reliability comes from U.S. Military Standards Handbook 217B (MIL-STD-217B, 1970) as quoted in (Carter, 1986). The failure discussion draws heavily from (Little, 1991). There are a number of formal treatments of reliability and the associated mathematics, including (Ebeling, 1997) and (Lewis, 1987). The definition of maintainability is from (Ebeling, 1997). The distinction between service, inspection, and repair is from (Pahl and Beitz, 1996).

Section 8.4: Codes of ethics for engineers are discussed further in Chapter 9. Sections 8.4.1 and 8.4.2 draw heavily on (Rubin, 2001), which also includes a very instructive example of LCA written by Cliff Davidson.

Section 8.5: The definition of quality is from (Juran, 1979). The standard reference to the house of quality is (Hauser and Clausing, 1988), and a number of modifications and extensions have been offered since. The generalized diagram is an adaptation based on (Ullman, 1997) who devotes an entire chapter to a more detailed methodology for developing a house of quality.

8.7 EXERCISES

8.1 If you were asked to design a product for recyclability, how would you determine what that meant? In addition, what sorts of questions should you be prepared to ask and answer?

8.2 How might DFA considerations differ for products made in large volume (e.g., the portable electric guitar) and those made in very small quantities (e.g., a greenhouse)?

8.3 Your design team has produced two alternative designs for city buses. Alternative *A* has an initial cost of $100,000, estimated annual operating costs of $10,000, will require a $50,000 overhaul after five years. Alternative *B* has an initial cost of $150,000, estimated annual operating costs of $5,000, will not require an overhaul after five years. Both alternatives will last ten years. If all other vehicle performance characteristics are the same, determine which bus is preferable using a discount rate of 10%.

8.4 Would the decision reached in Exercise 8.3 change if the discount rate was 20%? What would happen if, instead, the discount rate was 15%? How do the resulting cost figures influence your assessments of the given cost estimates?

8.5 Your design team has produced two alternative designs for greenhouses in a developing country. Alternative *A* has an initial cost of $200 and will last two years. Alternative *B* has an initial cost of $1,000 and will last ten years. All other things being equal, determine which greenhouse is more economical with a discount rate of 10%. Which other other factors can influence this decision?

8.6 What is the reliability of the system portrayed in Figure 8.2?

8.7 What is the reliability of the system portrayed in Figure 8.3? How does this result compare with that of Exercise 8.6? Why?

8.8 On what basis would you choose between a single system, all of whose parts are redundant, and two copies of a system that has no redundancy?

8.9 What are the factors to consider in an environmental analysis of the beverage container design problem? How might an environmental life cycle assessment help in addressing some of these questions?

8.10 Draw the house of quality for the beverage container design problem used in the earlier chapters.

CHAPTER 9

ETHICS IN DESIGN

Isn't design really just a technical matter?

IN **OUR** general discussions of design in Chapter 1, we saw that design is a very human endeavor. It involves the interactions among members of a design team, the relationships between designers, clients, and manufacturers, and the ways that purchasers of designed devices use them in their lives. Since design touches so many facets of people's daily lives, we must consider how people interact with each other as they create and use a design. To design means to accept responsibility for creating designs for people. That is, design is not done in a vacuum; design is a social activity. Therefore, ethics and ethical behavior must be considered in our examination of how designs are created and used.

9.1 ETHICS: RECONCILING CONFLICTING OBLIGATIONS

Words like ethics, morals, obligations, and duty carry baggage. Therefore, we start this discussion with some dictionary definitions. First, the word *ethics*:

> **ethics 1** the discipline dealing with what is good and bad and with moral duty and obligation **2 a:** a set of moral principles or values **b:** a theory or system of moral values **c:** the principles of conduct governing an individual or group

And since it is referenced so often in the definition of ethics, the word *moral*:

> **moral 1 a:** of or relating to principles of right or wrong in behavior **b:** expressing or teaching a conception of right behavior

Besides defining a discipline or field of study, these definitions define ethics as a set of guiding principles or a system people use to help them behave well. But didn't we learn right and wrong from our parents? Or perhaps we learned it as a set of belief in one of the religious traditions that emphasize faith in God (e.g., Christianity, Judaism, and Islam) or those that stress faith in a *right path* (e.g., Buddhism, Confucianism, and Taoism). Either way, we know about honesty and integrity, and about the injunction to treat others as we would want to be treated ourselves.

If we know these things, why do we need another, external set of rules to teach us what we already know? If we don't, and the law doesn't keep us in line, what is the use of

Ethics are
principles
of conduct
for both
individuals
and groups.

a set of ethical principles? The truth is that the kinds of lessons we learn at home, in school, and in our religious forums, may not provide enough explicit guidance about many of the situations we face in life, especially in our professional lives. In addition, given the diversity and complexity of our society, we are better off to have some standards of professional behavior that are universally agreed upon, across all of our traditions and individual upbringings. (Our dependence on laws and lawyers would be significantly and happily diminished if everyone's individually-learned lessons were sufficient!)

Our professional lives are complicated because many of our newfound responsibilities involve obligations to many stakeholders, some of whom are obvious (e.g., clients, users, the public) and some of whom are not (e.g., some government agencies, professional societies). We will elaborate these obligations further in Sections 9.2 through 9.4, but we note from the outset that these obligations often conflict. For example, a client may want one thing, while a group of people affected by a design may want something altogether different. Further, those affected by a design may not even know how they are being affected until *after* the design is complete and has been implemented.

Consider the following scenario: Imagine you are a mining engineer who has been engaged by the owner of a mine somewhere in the United States, to design a new shaft extension. As part of that design task you survey the mine and find that part of it runs under someone else's property. Are you obligated to simply complete the survey and the design for the mine owner who engaged you and who is paying you, and then go on to your next professional engagement?

You may suspect that the mine owner hasn't notified the landowner that his mineral rights are being excavated out from underneath him. Should you do something about that? If so, what? Further, what compels you to do something? Is it personal morality? Is there a law? How are you responsible, and to whom?

The chain of questions just started can easily be lengthened, and the situation made more complicated. For example, if the mine was the only mine in town and its owner controlled your livelihood, and those of many town residents. Or, if you find out that the mine runs under, or perilously close to, a school: does that change things?

This story highlights some of the many actors and obligations that could arise in an engineering design project. It is a bit like a case study because scenarios such as this occurred late in the 19th century and early in the 20th. It was just such situations that provided impetus to the formation of professional societies and the development of codes of ethics by these societies, in part as a form of protection for the individual members.

Over time, the professional societies also undertook other kinds of activities, including promulgating standards for design endeavors, such as those described in Section 8.3, and providing forums for reporting research and innovations in practice. But the main point is that professional engineering societies play an important role in setting ethical standards for designers and engineers. These ethical standards clearly speak to the various and often conflicting obligations that an engineer must meet. The societies also provide mechanisms for helping engineers deal with and resolve conflicting obligations, and, when asked, they provide the means for investigating and evaluating ethical behavior.

To return to our mining task, we can see that you might have to try to meet a range of obligations as the designer. You owe the mine owner a professional shaft design, including an unassailable survey of the mine itself. However, having discovered that the mine is under someone else's land, you might feel a need to ask the mine owner about what he's doing, approach the landowner and see what she knows, or go to some external authority either to ask questions or raise allegations. In the later version of this example, where a school may be above the mine, more people (i.e., schoolchildren and their parents) and agencies (i.e., the school board) would be involved, as a result of which you might feel still more obligations. In either scenario, as the designer you would likely feel pulled or obligated toward several people or groups. That is why there is a need for some structure or guidelines to assist engineers in framing a response and formulating a plan to deal with these kinds of concerns.

The situation described in our example is far from unique. One of the better-known instances of a difficult and noteworthy ethical dilemma was faced by the group of engineers that tried, unsuccessfully, to delay the launch of the space shuttle *Challenger* on January 28, 1986. While severe doubt was expressed by some engineers about the safety of the *Challenger*'s O-rings because of the cold weather before the flight, the upper management of Morton-Thiokol, the company that made the *Challenger*'s booster rockets, and NASA approved the launch. These managers determined that their concerns about Morton-Thiokol's image and the stature and visibility of NASA's shuttle program outweighed the judgments of the engineers closest to the booster design. The Morton-Thiokol engineers later publicized the dismissal of their recommendation not to launch by engaging in *whistleblowing*, wherein someone "blows the whistle" in order to stop a faulty decision made within a company, agency, or some other institution.

Whistleblowing is not new or unique. One famous early case is that of an industrial engineer, Ernest Fitzgerald, who blew the whistle on major cost overruns in the procurement of the Air Force's giant C–5A cargo plane. The Air Force was so displeased with Fitzgerald's actions that it took bureaucratic actions to keep him from further work on the C–5A: It "lost" his Civil Service tenure and then reconstructed that part of the bureaucracy in which Fitzgerald worked so as to eliminate his position! After an arduous and expensive legal battle, Fitzgerald earned a substantial settlement for wrongful termination and was reinstated to his position.

While such stories are to some extent discouraging, they also show heroic behavior under trying circumstances. More to the point, these examples show how "doing what's right" can be perceived quite differently within an organization. An engineer may well be faced with just the kind of clash of obligations that lies at the crux of any discussion of engineering ethics. If that happens, to whom can the designer or engineer turn to for help?

Part of the answer lies in the roots of the engineer's own understanding of her ethics, that is, in herself, her family, and her own personal beliefs. Another part of the answer lies in the support of immediate professional colleagues and peers, as they have been seen to be very effective in both righting the perceived wrong and in sustaining the whistleblower. And there are the professional societies.

Most professional engineering societies have published codes of ethics. We show the codes of ethics of the American Society of Civil Engineers (ASCE) in Figure 9.1 and of the Institute of Electronics and Electrical Engineers (IEEE) in Figure 9.2. While both codes emphasize integrity and honesty, they do appear to value certain kinds of behavior differently. For example, the ASCE code enjoins its members from competing unfairly with others, a subject not mentioned by the IEEE. Similarly, the IEEE specifically calls for its members to "fairly treat all persons regardless of such factors as race, religion, gender . . ."

ASCE CODE OF ETHICS

<u>Fundamental Principles</u>

Engineers uphold and advance the integrity, honor and dignity of the engineering profession by:

1. using their knowledge and skill for the enhancement of human welfare and the environment;

2. being honest and impartial and serving with fidelity the public, their employers and clients;

3. striving to increase the competence and prestige of the engineering profession; and

4. supporting the professional and technical societies of their disciplines.

<u>Fundamental Canons</u>

1. Engineers shall hold paramount the safety, health and welfare of the public and shall strive to comply with the principles of sustainable development in the performance of their professional duties.

2. Engineers shall perform services only in areas of their competence.

3. Engineers shall issue public statements only in an objective and truthful manner.

4. Engineers shall act in professional matters for each employer or client as faithful agents or trustees, and shall avoid conflicts of interest.

5. Engineers shall build their professional reputation on the merit of their services and shall not compete unfairly with others.

6. Engineers shall act in such a manner as to uphold and enhance the honor, integrity, and dignity of the engineering profession.

7. Engineers shall continue their professional development throughout their careers, and shall provide opportunities for the professional development of those engineers under their supervision.

FIGURE 9.1 The code of ethics of the American Society of Civil Engineers (ASCE), dated January 1, 1977. It is similar, although not identical, to the code adopted by the IEEE that is displayed in Figure 9.2.

There are other differences as well, for example, in the styles of language. The ASCE presents a set of injunctions about what engineers "shall" do, while the IEEE code is phrased as a set of commitments to undertake certain behaviors.

One of an engineer's obligations is to adhere to a code of ethics.

Notwithstanding these differences, both codes of ethics set out guidelines or standards of how to behave with respect to: clients (e.g., ASCE's "as faithful agents or trustees"); the profession (e.g., IEEE's "assist colleagues and co-workers in their professional development"); the law (e.g., IEEE's "reject bribery in all its forms"); and the public (e.g., ASCE's "shall issue public statements only in an objective and truthful manner"). These standards lay out rules of the road for dealing with conflicting obligations, including the task of assessing whether these conflicts are "only" of perception or of a "real" and potentially damaging nature.

IEEE CODE OF ETHICS

We, the members of the IEEE, in recognition of the importance of our technologies in affecting the quality of life throughout the world, and in accepting a personal obligation to our profession, its members and the communities we serve, do hereby commit ourselves to the highest ethical and professional conduct and agree:

1. to accept responsibility in making engineering decisions consistent with the safety, health and welfare of the public, and to disclose promptly factors that might endanger the public or the environment;

2. to avoid real or perceived conflicts of interest whenever possible, and to disclose them to affected parties when they do exist;

3. to be honest and realistic in stating claims or estimates based on available data;

4. to reject bribery in all its forms;

5. to improve the understanding of technology, its appropriate application, and potential consequences;

6. to maintain and improve our technical competence and to undertake technological tasks for others only if qualified by training or experience, or after full disclosure of pertinent limitations;

7. to seek, accept, and offer honest criticism of technical work, to acknowledge and correct errors, and to credit properly the contributions of others;

8. to treat fairly all persons regardless of such factors as race, religion, gender, disability, age, or national origin;

9. to avoid injuring others, their property, reputation, or employment by false or malicious action;

10. to assist colleagues and co-workers in their professional development and to support them in following this code of ethics.

FIGURE 9.2 The code of ethics of the Institute of Electronics and Electrical Engineers (IEEE), dated August 1990. How does the IEEE code of ethics differ from that adopted by the ASCE that is displayed in Figure 9.1?

We close with two more notes about professional societies and their codes. First, the differences in the codes reflect different styles of engineering practice in the various disciplines much more than differences in their views of the importance of ethics. For example, most civil engineers not employed by a government agency work in small companies that are people-intensive, rather than capital-intensive. These firms obtain much of their work through public, competitive bidding. Electrical engineers, on the other hand, more often than not work for large corporations that sell products more than services, one result of which is that they have significant manufacturing operations and are capital-intensive. Such different practices produce different cultures and, hence, different statements of ethical standards.

The second point is that the professional societies have not always been seen as active and visible protectors of whistleblowers and other professionals who raise concerns about specific engineering or design instances, notwithstanding their promulgation of codes of ethics. This situation is improving, steadily if slowly, but many engineers still find it difficult to look to their societies, and especially their local branch sections, for first-line assistance and support in times of need. Of course, as we all make ethical behavior a higher priority, the need for such support will lessen, and its ready availability will surely increase.

9.2 OBLIGATIONS MAY START WITH THE CLIENT . . .

Consider now our various obligations to a client or to an employer. As designers or engineers, we owe our client or employer a professional effort at solving a design problem, by which we mean being technically competent, conscientious and thorough, and that we should undertake technical tasks only if we are properly "qualified by training or experience." We must avoid any conflicts of interest, and disclose any that may exist. And we must serve our employer by being "honest and impartial" and by "serving with fidelity . . . " Most of these obligations are clearly delineated in codes of ethics (e.g., compare the quotes with Figures 9.1 and 9.2), but there's at least one curious obligation on this list: What does it mean to serve with "fidelity"?

A thesaurus would tell us that fidelity has several synonyms, including constancy, fealty, allegiance, and loyalty. Thus, one implication we can draw from the ASCE code of ethics is that we should be loyal to our employer or client. This suggests that one of our obligations is to look out for the best interests of our client or employer, and to maintain a clear picture of those interests as we do our design work. But loyalty is a very touchy matter; it is not a simple, one-dimensional attribute. In fact, clients and companies earn the loyalty of their consultants and staffs in at least two ways. One, called *agency-loyalty*, derives from the nature of any contracts between the designer and his client (e.g., "work for hire") or his employer (e.g., a "hired worker"). As it is dictated by contract, agency-loyalty is clearly obligatory for the designer. The second kind of loyalty, *identification-loyalty*, is more likely to be seen as optional. It stems from the engineer identifying with the client or company because he admires its goals or sees its behavior as mirroring his own values. To the extent that identification-loyalty is optional, it will be earned by clients and companies only if they demonstrate reciprocal loyalty to their staff designers.

Agency-loyalty provides one reason to maintain a "design notebook" to document design work. As we have noted before, keeping such a record is good design practice because it is very useful for recapitulating our thinking as we move through different stages of the design process and for real-time tracking. A dated design notebook also provides a legal basis for documenting how new, patentable ideas were developed. Such documentation is essential to an employer or client if a patent application is in any way challenged. Further, as is typically specified in contracts and employment agreements, the intellectual work done in the creation of a design is itself the intellectual property of the client or the employer. A client or an employer may share the rights to that intellectual property with its creators, but the fundamental decisions about the ownership of the property generally belong to the client or owner. It is important for a designer to keep that in mind, and also to document any separate, private work that she is doing, just to avoid any confusion about who owns any particular piece of design work.

Since identification-loyalty is optional, it provides fertile ground for clashes of obligations because other loyalties have the space to make themselves felt here. As we will dis-

Engineers must be loyal to their clients, to other stakeholders, and to themselves.

cuss further in Section 9.3.1, modern codes of ethics normally articulate some form of obligation to the health and welfare of the public. For example, the ASCE code of ethics (see Figure 9.1) suggests that civil engineers work toward enhancing both human welfare and the environment, and that they "shall hold paramount the safety, health, and welfare of the public . . . " Similarly, the IEEE code (see Figure 9.2) suggests that its members commit to "making engineering decisions consistent with the safety, health and welfare of the public . . . " These are clear calls to engineers to identify other loyalties to which they should feel allegiance. There is little doubt that it was just such divided loyalties that emerged in the many reported cases of whistleblowing.

In the case of the explosion of the *Challenger*, those who argued against its launching felt that lives would be endangered. They placed a higher value on the lives being risked than they did on the loyalties demanded by Morton-Thiokol (i.e., to secure its place as a government contractor) and by NASA (i.e., to its ability to successfully argue for the shuttle program before Congress and the public). Similar conflicting loyalties emerge for engineers as toxic waste sites are cleaned up under the Super Fund program of the Environmental Protection Agency (EPA). In many instances, employees felt that they needed to look out for their own companies, sometimes because they felt the companies should not be penalized for doing what was once legal, in others because they were pressured to do so by peers and bosses and the possible loss of their jobs. There are also cases in which engineers were apparently willing, or at least able, to rank their loyalties to their companies first, to the point where falsified emission test data were reported to the government (by engineers and managers at the Ford Motor Company) or parts known to be faulty were delivered to the Air Force (by engineers and managers at the B. F. Goodrich Company).

An apparent disloyalty to a company or an organization may sometimes be, in a longer term, an act of greater, successfully merged loyalties. When the Ford Pinto was initially being designed, for example, some of its engineers wanted to perform crash tests that were not then called for in the relevant U.S. Department of Transportation regulations. The managers charged with developing the Pinto felt that such tests could not benefit the program and, in fact, might only prove to be a burden. Why run a test that is not required, only to risk failing that test? The designers who proposed the tests were seen as disloyal to Ford and to the Pinto program. In fact, the placement of the drive train and the gas tank resulted in fiery crashes, lives lost, and major public relations and financial headaches for Ford. Clearly, Ford would have been better off in the long run to have conducted the tests, so the engineers who proposed them could be said to have been looking out for the company's long-term interests.

If there is one point that emerges from the discussion thus far, it is that ethical issues do not arise from a *single* obligation. Indeed, were issues so easily categorized, choices would vanish and ethics would not be a problem.

9.3 . . . BUT WHAT ABOUT THE PUBLIC AND THE PROFESSION?

We now tell a story that shows that when people behave well, things can turn out well even in bad situations. In fact, to start with the ending, the protagonist-hero of this story said, "In return for getting a [professional engineering] license and being regarded with respect, you're supposed to be self-sacrificing and look beyond the interests of yourself and your client to society as a whole. And the most wonderful part of my story is that when I did it nothing bad happened."

Our hero is William J. LeMessurier (pronouncd "LeMeasure") of Cambridge, Massachusetts, one of the most highly regarded structural engineers and designers in the world. He served as the structural consultant to a noted architect, Hugh Stubbins, Jr., for the design of a new New York headquarters for Citicorp, the parent corporation of one of the nation's largest banks, CitiBank. Completed in 1978, the 59-story Citicorp Center is one of the most dramatic and interesting skyscrapers in a city filled with some of the world's great buildings (see Figure 9.3). In many ways, LeMessurier's conceptual design for Citicorp resembles other striking skyscrapers in that it used the *tube* concept in which a building is designed as a tall, hollow tube that has a comparatively rigid or stiff tube wall. (In structural engineering terminology, the tube's main lateral stability elements are located at the outer perimeter and tied together at the corners.) Fazlur Kahn's John Hancock Center in Chicago is a similar design (see Figure 9.4). The outer "tube" or "main lateral stability elements" are the multistory diagonal elements that are joined to large columns at the corners. Kahn's design benefited from a deliberate architectural decision to expose the tube's details, perhaps to illustrate the famous dictum that *form follows function* (which is routinely and wrongly attributed to Frank Lloyd Wright, but was pronounced by Louis Sullivan, the noted Chicago architect who was also Wright's mentor).

LeMessurier's Citicorp design was innovative in several ways. One, not visible from the outside, was the inclusion of a large mass, floating on a sheet of oil, within the triangular roof structure. It was added as a damper to reduce or damp out any oscillations the building might undergo due to wind forces. Another innovation was LeMessurier's adaptation of

FIGURE 9.3 One view of the 59-story Citicorp Center, designed by architect Hugh Stubbins, Jr., with William J. LeMessurier serving as structural consultant. One of this building's notable features is that it rests on four massive columns that are placed at the midpoints of the building's sides, rather than at the corners. This enabled the architects to include under the Citicorp's sheltering canopy a new building for St. Peter's Church. (Photo by Clive L. Dym.)

FIGURE 9.4 The 102-story John Hancock Center, designed by the architectural firm of Skidmore, Owings and Merrill, with Fazlur Kahn serving as structural engineer. Note how the exposed diagonal and column elements make up the tube that is the building's underlying conceptual design. (Photo by Clive L. Dym.)

the tube concept to an unusual situation. The land on which the Citicorp Center was built had belonged to St. Peter's Church, with the church occupying an old (dating from 1905) and decaying Gothic building on the lot's west side. When St. Peter's sold the building lot to Citicorp, it also negotiated that a new church be erected "under" the Citicorp skyscraper. In order to manage this, LeMessurier moved the "corners" of the building to the midpoints of each side (see Figure 9.5). This enabled the creation of a large space for the new church because the office tower itself was then cantilevered out over the church, at a height of some nine stories. Looking at the sidewalls of the tube—and here we have to peel off the building's skin because architect Stubbins did not want the structures exposed, as they were in the Hancock tower—we see that the wall's rigidity comes from large triangles, made up of diagonal and horizontal elements, that all connect at the midpoints of the sides. Thus, LeMessurier's triangles serve the same purpose as Kahn's large X-frames.

The ethics problem arose soon after the building was completed and occupied. LeMessurier received a call from an engineering student in New Jersey who was told by a professor that the building's columns had been put in the wrong place. In fact, LeMessurier

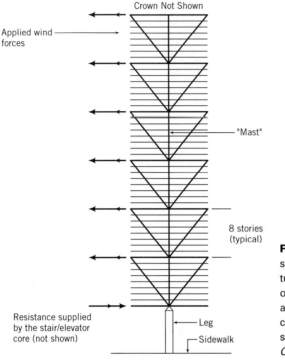

Crown Not Shown

Applied wind forces

"Mast"

8 stories (typical)

Resistance supplied by the stair/elevator core (not shown)

Leg

Sidewalk

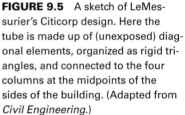

FIGURE 9.5 A sketch of LeMessurier's Citicorp design. Here the tube is made up of (unexposed) diagonal elements, organized as rigid triangles, and connected to the four columns at the midpoints of the sides of the building. (Adapted from *Civil Engineering.*)

was very proud of his idea of placing the columns at the midpoints. He explained to the student how the 48 diagonal braces he had superposed on the midspan columns added great stiffness to the building's tube framework, particularly with respect to wind forces. The student's questions sufficiently intrigued LeMessurier that he reviewed his original design and calculations to see just how strong the wind bracing system would be. He found himself looking at a case that was not examined under then-current practice and building codes. Practice at the time called for wind force effects to be calculated when the wind flow hit a side of a building dead on, that is, normal to the building faces. However, the calculation of the effect of a *quartering wind*, under which the wind hits a building on a 45-degree diagonal and the resulting wind pressure is then distributed over the two immediately adjacent faces (see Figure 9.6), had not been called for previously. A quartering wind on the Citicorp Center leaves some diagonals unstressed and others doubly loaded, with calculated strain increases of 40 percent. Normally, even this increase in strain (and stress) would not have been a problem because of the basic assumptions under which the entire system was designed.

However, to LeMessurier's further discomfort, he found out just a few weeks later that the actual connections in the finished diagonal bracing system were not the high-strength welds he had stipulated. Rather, the connections were bolted because Bethlehem Steel, the steel fabricator, had determined and suggested to LeMessurier's New York office that bolts would be more than strong enough and, at the same time, significantly cheaper. The choice of bolts was sound and quite correct professionally. However, to LeMessurier the bolts meant that the margin of safety against forces due to a quartering wind—which,

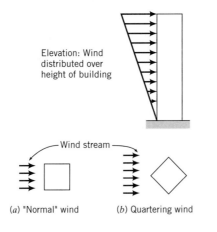

Elevation: Wind distributed over height of building

Wind stream

(a) "Normal" wind   (b) Quartering wind

FIGURE 9.6   A sketch of how wind forces are seen on buildings. (a) This is the standard case of wind streams being normal or perpendicular to the building faces. It was the only one called for in design codes at the time the Citicorp center was designed. (b) This is the case of quartering winds, in which case the wind stream come along a 45-degree diagonal and thus simultaneously applies pressure along two faces at once.

again, structural engineers were not then called upon to consider—was not as large as he would have liked. (It is interesting to note that while New York City's building code did not then require that quartering winds be considered in building design, Boston's code did— and had since the 1950s!)

Spurred by his new calculations and the news about the bolts, and by hearing of some other detailed design assumptions made by engineers in his New York office, LeMessurier retreated to the privacy of his summer home on an island in Maine to carefully review all of the calculations and changes, and their implications. After doing a member-by-member calculation of the forces and reviewing the weather statistics for New York City, LeMessurier determined that once every sixteen years, on average, the Citicorp Center would be subjected to winds that could produce a catastrophic failure. Thus, in terminology used by meteorologists to describe both winds and floods, the Citicorp Center would fail in a sixteen-year storm—when it had supposedly been designed to withstand a fifty-year storm. So, what was LeMessurier to do?

In fact, LeMessurier considered several options, reportedly including driving into a freeway bridge abutment at high speed. He also considered remaining silent, as he tried to reassure himself that his innovative rooftop mass damper actually reduced the probabilities of such failure to the fifty-year level. On the other hand, if the power went out, the mass damper wouldn't be there to help. So, what did LeMessurier actually do?

He first tried to contact the architect, Hugh Stubbins, who was away on a trip. He then called Stubbins' lawyer, after which he talked first with his own insurance carrier and then with the principal officers of Citicorp, one of whom studied engineering before choosing to become a banker. While some early consideration was given to evacuating the building, especially since hurricane season was just over the horizon, it was decided instead that all of the connections at risk should be redesigned and retroactively fixed. Steel plate "Band-Aids," two inches thick, would be welded onto each of 200 bolted connections. However, there were some very interesting implementation problems, only some of which we mention here. The building's occupants had to be informed without being alarmed, because the repair work would go on every night for two months or more. The public had to be informed why the bank's new flagship headquarters suddenly needed immediate modifications. (In fact, the entire process was open and attentive to public concerns.) Skilled structural welders, who were in short supply, had to be found, as did an adequate supply of the right grade of steel plate. Discrete, even secret, evacuation plans had to be put in place, just in case an unexpectedly high wind did show up while the repairs were being made. And New York City's Building Commissioner and the Department of Buildings and its inspectors had to be brought into the loop because they were central to resolving the problem. They had to be informed about the problem and its proposed solution, and they had to agree to inspect that solution. All in all, a dazzling array of concerns, institutions, and, of course, personalities.

In the end, the steel Band-Aids were applied and the entire business was completed professionally, with no finger-pointing, no public assignments of blame. LeMessurier, who had thought his career might precipitously end, came away with still greater stature, occasioned by his willingness to face up to the problem candidly and propose a realistic, carefully crafted solution. In the words of one of the engineers involved in implementing LeMessurier's solution, "It wasn't a case of 'We caught you, you skunk.' It started with a guy who stood up and said, 'I got a problem, I made the problem, let's fix the problem.' If you're gonna kill a guy like LeMessurier, why should anybody ever talk?"

As we have said before, this is a case where everyone involved behaved well. In fact, to everyone's credit, all of the actors behaved to a very high standard of professionalism and understanding. It is, therefore, a case that we can study with pleasure, particularly as engineers. It also is a case that could have gone other ways, so we will close our discussion by posing a few questions that you, the reader, might face were you in LeMessurier's position:

- Would you have "blown the whistle" or not?

- What would you have done if you determined that the revised probability of failure was higher (i.e., worse) than for the original design, but still within the range permitted by code?

- What would you have done if your insurance carrier had said to "keep quiet?"

- What would you have done if the building owner, or the city, had said to "keep quiet?"

- Who should pay for the repair?

9.4   IS IT OK FOR ME TO BE WORKING ON THIS PROJECT?

The cases we have mentioned seem to be more about engineering practice than about design *per se*. But, in fact, the cases are about design as a social activity because the engineers involved were clearly working to realize designs they had created. Perhaps a more telling point arises when we are asked to design a product that we think needn't be made, or even shouldn't be made. In Chapter 4 we referred to the design of a cigarette lighter, which we also thought of as an igniter of leafy matter. While this example seems trivial, even nonsensical, it points to another facet of divided loyalties. It suggests that designing cigarette lighters is, somehow, morally troubling. In the United States today, it would seem that designing cigarette lighters and cigarette-making machinery might be viewed as at least politically incorrect, and therefore morally incorrect, by a large fraction of the population. On the other hand, isn't it up to individuals to choose to smoke or not? If a product is legal, shouldn't we allow ourselves to design it without feeling uncomfortable?

Not everything that can be designed should be designed.

A more glaring instance along these lines arises if we look at the designers of the large scale ovens and their specialized buildings made in Germany in the 1930s and 1940s, and to some, the nuclear weapons designed both here and in the Soviet Union since the late 1940s (and today in a number of developing countries!) The technologies were different, and while some engineers and physicists were excited by the intellectual challenge of designing devices to harness nuclear fission, it's hard to imagine oven designers having similar feelings. So, were these two sets of designers merely being loyal to their clients, to their governments, and their societies? If so, were they being loyal toward "human welfare and the environment?"

This is an issue that is intensely personal, that is, should *I* be working on this design project? Unfortunately, there is no way to know in advance when a serious conflict of obligations and loyalties will arise in any of our lives. Nor can we know the specific personal and professional circumstances within which that conflict will be embedded. Nor, unfortunately, is there a single answer to some of the questions posed, beyond, "It depends." If faced with a daunting conflict, we can only hope that we are prepared by our upbringing, our maturity, and our ability to think and reflect about the issues that we have so briefly raised herein.

9.5 NOTES

Section 9.1: (Martin and Schinzinger, 1996) and (Glazer and Glazer, 1989) are very interesting, useful, and readable books on, respectively, engineering ethics and whistleblowing. Ethics emerges as a major theme in Harr's tale (1995) of a civil lawsuit spawned by inadequate toxic waste cleanup.

Section 9.2: The definitions of agency- and identification-loyalty are derived from (Martin and Schinzinger 1996).

Section 9.3: The Citicorp Center case is adapted from (Morgenstern, 1995) and (Goldstein and Rubin, 1996). We were greatly helped by William LeMessurier's review of the material.

Section 9.4: For the sake of brevity, we cite only (Arendt, 1963) and (Harr, 1995) among the many books about the deep and complex issues raised in this section.

9.6 EXERCISES

9.1 Is there a difference between ethics and morals?

9.2 Identify the stakeholders that the design team at HMCI must recognize as it develops its design for the portable electric guitar. Are there obligations to these stakeholders that you should consider and could they conflict with what your client has asked you to do?

9.3 As an engineer testing designs for electronic components, you discover that they fail in a particular location. Subsequent investigation shows that the failures are due to a nearby high-powered radar facility. While you can shield your own designs so that they will work in this environment, you also notice that there is an adjacent nursery school. What actions, if any, should you take?

9.4 You are considering a safety test for a newly-designed device. Your supervisor instructs you not to perform this test because the relevant government regulations are silent on this aspect of the design. What actions, if any, should you take?

9.5 As a result of previous experiences as a designer of electronic packaging, you understand a sophisticated heat-treatment process that has not been patented, although it is considered company confidential. In a new job, you are designing beverage containers for BJIC and you believe that this heat treatment process could be effectively used. Can you use your prior knowledge?

9.6   With reference to Exercise 9.5, suppose your employer is a nonprofit organization that is committed to supplying food to disaster victims. Would that change the actions you might take?

9.7   You are asked to provide a reference for a member of your design team, Jim, in connection with a job application he has filed. You have not been happy with Jim's performance, but you believe that he might do better in a different setting. While you are hopeful that you can replace Jim, you also feel obligated to provide an honest appraisal of Jim's potential. What should you do?

9.8   With reference to Exercise 9.7, would your answer change if you knew that you could not replace Jim?

BIBLIOGRAPHY

In addition to providing the references cited in the text, the bibliography below is a sampling of books covering a wide range of issues including design theory, design in different disciplines, project management techniques, optimization theory, some applications of artificial intelligence, engineering ethics, and the practice of engineering, and more. This list of works is *not* complete—the literatures on design and project management alone are both vast and rapidly expanding. Thus, it should be kept in mind that this bibliography represents only the tip of a very large iceberg of published work in design and in project management. Some of the works cited are just intellectually interesting, and some are books that students in particular will find useful for project work.

J. L. Adams, *Conceptual Blockbusting: A Guide to Better Ideas*, Stanford Alumni Association, Stanford, CA, 1979.

K. Akiyama, *Function Analysis: Systematic Improvement of Quality and Performance*, Productivity Press, Cambridge, MA, 1991.

C. Alexander, *Notes on the Synthesis of Form*, Harvard University Press, Cambridge, MA, 1964.

Anon., *Goals and Priorities for Research in Engineering Design*, American Society of Mechanical Engineers, New York, NY, 1986.

Anon., *Improving Engineering Design: Designing for Competitive Advantage*, National Research Council, National Academy Press, Washington, D.C., 1991.

Anon., *Managing Projects and Programs*, The Harvard Business Review Book Series, Harvard Business School Press, Cambridge, MA, 1989.

E. K. Antonsson and J. Cagan, *Formal Engineering Design Synthesis*, Cambridge University Press, New York, 2001.

H. Arendt, *Eichmann in Jerusalem: A Report on the Banality of Evil*, Viking Press, New York, NY, 1963.

J. S. Arora, *Introduction to Optimum Design*, McGraw-Hill, New York, NY, 1989.

K. J. Arrow, *Social Choice and Individual Values*. John Wiley & Sons, Inc., New York, 1951.

W. Asimow, *Introduction to Design*, Prentice-Hall, Englewood Cliffs, NJ, 1962.

A. B. Badiru, *Project Management in Manufacturing and High Technology Operations*, John Wiley & Sons, Inc., New York, NY, 1996.

K. M. Bartol and D. C. Martin, *Management*, 2nd edit., McGraw-Hill Book Company, New York, NY, 1994.

Barton-Aschman Associates, *North Area Terminal Study*, Technical Report, Barton-Aschman Associates, Evanston, IL, August 1962.

Louis Berger, *Central Artery North Area Project*, Interim Report, Louis Berger & Associates, Cambridge, MA, 1981.

G. Boothroyd and P. Dewhurst, *Product Design for Assembly*, Boothroyd Dewhurst Inc., Wakefield, RI, 1989.

T. Both, G. Breed, C. Stratton and K. V. Horn, *Micro Laryngeal Surgery: An Instrument Stabilizer*, E4 Project Report, Department of Engineering, Harvey Mudd College, Claremont, CA, 2000.

C. L. Bovee, M. J. Houston and J. V. Thill, *Marketing*, 2nd edit., McGraw-Hill Book Company, New York, NY, 1995.

C. L. Bovee, J. V. Thill, M. B. Word and G. P. Dovel, *Management*, McGraw-Hill Book Company, New York, NY, 1993.

E. T. Boyer, F. D. Meyers, F. M. Croft, Jr., M. J. Miller and J. T. Demel, *Technical Graphics*, John Wiley & Sons, Inc., New York, NY, 1991.

D. C. Brown, "Design," in S. C. Shapiro (Editor), *Encyclopedia of Artificial Intelligence*, 2nd Edition, John Wiley & Sons, Inc., New York, NY, 1992.

D. C. Brown and B. Chandrasekaran, *Design Problem Solving*, Pitman, London, and Morgan Kaufmann, Los Altos, CA, 1989.

L. L. Bucciarelli, *Designing Engineers*, MIT Press, Cambridge, MA, 1994.

S. Carlson Skalak, H. Kemser and N. Ter-Minassian, "Defining a Product development Methodology with Concurrent Engineering for Small Manufacturing Companies," *Journal of Engineering Design*, 8 (4), 305–328, December 1997

A. D. S. Carter, *Mechanical Reliability*, Macmillan, London, England, 1986.

S. Chan, R. Ellis, M. Hanada and J. Hsu, *Stabilization of Microlaryngeal Surgical Instruments*, E4 Project Report, Department of Engineering, Harvey Mudd College, Claremont, CA, 2000.

J. Connor, K. Kubler, P. Leitzell, J. P. Strozzo and M. Wang, *Design of a Chicken Coop*, E4 Project Report, Department of Engineering, Harvey Mudd College, Claremont, CA, 1997.

J. Corbett, M. Dooner, J. Meleka and C. Pym, *Design for Manufacture: Strategies, Principles and Techniques*, Addison-Wesley, Wokingham, England, 1991.

R. D. Coyne, M. A. Rosenman, A. D. Radford, M. Balachandran and J. S. Gero, *Knowledge-Based Design Systems*, Addison-Wesley, Reading, MA, 1990.

N. Cross, *Engineering Design Methods*, 2nd edit., John Wiley & Sons, Chichester, England, 1994.

M. L. Dertouzos, R. K. Lester, R. M. Solow and the MIT Commission on Industrial Productivity, *The Making of America: Regaining the Productive Edge*, MIT Press, Cambridge, MA, 1989.

J. R. Dixon, *Design Engineering: Inventiveness, Analysis, and Decision Making*, McGraw-Hill, New York, NY, 1966.

J. R. Dixon, "Engineering Design Science: The State of Education," *Mechanical Engineering*, 113 (2), February 1991.

J. R. Dixon, "Engineering Design Science: New Goals for Education," *Mechanical Engineering*, 113 (3), March 1991.

J. R. Dixon and C. Poli, *Engineering Design and Design for Manufacturing*, Field Stone Publishers, Conway, MA, 1995.

C. L. Dym (Editor), *Applications of Knowledge-Based Systems to Engineering Analysis and Design*, American Society of Mechanical Engineers, New York, NY, 1985.

C. L. Dym (Editor), *Computing Futures in Engineering Design*, Harvey Mudd College, Claremont, CA, 1997.

C. L. Dym (Editor), *Designing Design Education for the 21st Century*, Harvey Mudd College, Claremont, CA, 1999.

C. L. Dym, *E4 (Engineering Projects) Handbook*, Department of Engineering, Harvey Mudd College, Claremont, CA, Spring 1993.

C. L. Dym, *Engineering Design: A Synthesis of Views*, Cambridge University Press, New York, NY, 1994a.

C. L. Dym, Letter to the Editor, *Mechanical Engineering*, 114 (8), August 1992.

C. L. Dym, "The Role of Symbolic Representation in Engineering Education," *IEEE Transactions on Education*, 35 (2), March 1993.

C. L. Dym, "Teaching Design to Freshmen: Style and Content," *Journal of Engineering Education*, 83 (4), 303–310, October 1994b.

C. L. Dym and E. S. Ivey, *Principles of Mathematical Modeling*, Academic Press, New York, NY, 1980.

C. L. Dym and R. E. Levitt, *Knowledge-Based Systems in Engineering*, McGraw-Hill, New York, NY, 1991.

C. L. Dym and L. Winner (Editors), *Social Dimensions of Engineering Design*, Harvey Mudd College, Claremont, CA, 2001.

C. L. Dym, W. H. Wood and M. J. Scott, "Rank Ordering Engineering Designs: Pairwise Comparison Charts and Borda Counts," *Research in Engineering Design*, 13 (4), 236–242, 2002.

C. E. Ebeling, *An Introduction to Reliability and Maintainability Engineering*, McGraw-Hill, New York, NY, 1997.

D. L. Edel, Jr., (Editor), *Introduction to Creative Design*, Prentice-Hall, Englewood Cliffs, NJ, 1967.

K. S. Edwards, Jr., and R. B. McKee, *Fundamentals of Mechanical Component Design*, McGraw-Hill, New York, NY, 1991.

K. A. Ericsson and H. A. Simon, *Protocol Analysis: Verbal Reports as Data*, MIT Press, Cambridge, MA, 1984.

A. Ertas and J. C. Jones, *The Engineering Design Process*, John Wiley & Sons, Inc., New York, NY, 1993.

D. L. Evans (Coordinator), "Special Issue: Integrating Design Throughout the Curriculum," *Engineering Education, 80* (5), 1990.

J. H. Faupel, *Engineering Design*, John Wiley & Sons, Inc., New York, NY, 1964.

L. Feagan, T. Galvani, S. Kelley and M. Ong, *Device for Microlaryngeal Instrument Stabilization*, E4 Project Report, Department of Engineering, Harvey Mudd College, Claremont, CA, 2000.

J. Fortune and G. Peters, *Learning From Failure—The Systems Approach*, John Wiley & Sons, Chichester, UK, 1995.

R. L. Fox, *Optimization Methods for Engineering Design*, Addison-Wesley, Reading, MA, 1971.

M. E. French, *Conceptual Design for Engineers*, 2nd Edition, Design Council Books, London, England, 1985.

M. E. French, *Form, Structure and Mechanism*, MacMillan, London, England, 1992.

D. C. Gause and G. M. Weinberg, *Exploring Requirements: Quality Before Design,* Dorset House Publishing, New York, NY, 1989.

J. S. Gero (Editor), *Design Optimization*, Academic Press, Orlando, FL, 1985.

J. S. Gero (Editor), *Proceedings of AI in Design '92*, Kluwer Academic Publishers, Dordrecht, The Netherlands, 1992.

J. S. Gero (Editor), *Proceedings of AI in Design '94*, Kluwer Academic Publishers, Dordrecht, The Netherlands, 1994.

J. S. Gero (Editor), *Proceedings of AI in Design '96*, Kluwer Academic Publishers, Dordrecht, The Netherlands, 1996.

M. P. Glazer and P. M. Glazer, *The Whistleblowers: Exposing Corruption in Government and Industry*, Basic Books, New York, NY, 1989.

G. L. Glegg, *The Design of Design*, Cambridge University Press, Cambridge, England, 1969.

G. L. Glegg, *The Science of Design*, Cambridge University Press, Cambridge, England, 1973.

G. L. Glegg, *The Selection of Design*, Cambridge University Press, Cambridge, England, 1972.

T. J. Glover, *Pocket Ref*, Sequoia Publishing, Littleton, CO, 1993.

S. H. Goldstein and R. A. Rubin, "Engineering Ethics," *Civil Engineering*, October 1996.

P. Graham (Editor), *Mary Parker Follett—Prophet of Management: A Celebration of Writings From the 1920s*, Harvard Business School Press, Boston, MA, 1996.

P. Gutierrez, J. Kimball, B. Maul, A. Thurston and J. Walker, *Design of a Chicken Coop*, E4 Project Report, Department of Engineering, Harvey Mudd College, Claremont, CA, 1997.

C. Hales, *Managing Engineering Design*, Longman Scientific & Technical, Harlow, England, 1993.

J. Harr, *A Civil Action*, Vintage Books, New York, NY, 1995.

B. Hartmann, B. Hulse, S. Jayaweera, A. Lamb, B. Massey and R. Minneman, *Design of a "Building Block" Analog Computer*, E4 Project Report, Department of Engineering, Harvey Mudd College, Claremont, CA, 1993.

J. R. Hauser and D. Clausing, "The House of Quality," *Harvard Business Review*, 63–73, May-June 1988.

S. I. Hayakawa, *Language in Thought and Action*, 4th Edition, Harcourt Brace Jovanovich, San Diego, CA, 1978.

R. T. Hays, "Value Management," in W. K. Hodson (Editor), *Maynard's Industrial Engineering Handbook*, 4th Edition, McGraw-Hill Book Company, New York, NY, 1992.

G. H. Hazelrigg, *Systems Engineering: An Approach to Information-Based Design.* Prentice Hall, Upper Saddle River, NJ, 1996.

G. H. Hazelrigg, "Validation of Engineering Design Alternative Selection Methods," unpublished manuscript, courtesy of the author, 2001

J. Heskett, *Industrial Design*, Thames and Hudson, London, 1980.

R. S. House, *The Human Side of Project Management*, Addison-Wesley, Reading, MA, 1988.

V. Hubka, M. M. Andreasen, and W. E. Eder, *Practical Studies in Systematic Design*, Butterworths, London, England, 1988.

B. Hyman, *Topics in Engineering Design*, Prentice Hall, Englewood Cliffs, NJ, 1998.

D. Jain, G. P. Luth, H. Krawinkler and K. H. Law, *A Formal Approach to Automating Conceptual Structural Design*, Technical Report No. 31, Center for Integrated Facility Engineering, Stanford University, Stanford, CA, 1990.

F. D. Jones, *Ingenious Mechanisms*: Vols. 1–3, The Industrial Press, New York, NY, 1930.

J. C. Jones, *Design Methods*, Wiley-Interscience, Chichester, UK, 1992.

J. Juran, *Quality Control Handbook*, 3rd Edition, McGraw-Hill, New York 1979.

D. Kaminski, "A Method to Avoid the Madness," *The New York Times*, 3 November 1996.

H. Kerzner, *Project Management: A Systems Approach to Planning, Scheduling and Controlling*, Van Nostrand Reinhold, New York, NY, 1992.

D. S. Kezsbom, D. L. Schilling and K. A. Edward, *Dynamic Project Management: A Practical Guide for Managers & Scientists*, John Wiley & Sons, Inc., New York, NY, 1989.

A. Kusiak, *Engineering Design: Products, Processes and Systems*, Academic Press, San Diego, CA, 1999.

M. Levy and M. Salvadori, *Why Buildings Fall Down*, Norton, New York, NY, 1992.

E. E. Lewis, *Introduction to Reliability Engineering*, John Wiley & Sons, Inc., New York, NY, 1987.

P. Little, *Improving Railroad Car Reliability Using A New Opportunistic Maintenance Heuristic and Other Information System Improvements*, Doctoral Dissertation, Massachusetts Institute of Technology, Cambridge, MA, 1991.

M. W. Martin and R. Schinzinger, *Ethics in Engineering*, 3rd Edition, McGraw-Hill Book Company, New York, NY, 1996.

Massachusetts Department of Public Works, *North Terminal*, Draft Environmental Impact Report (Section 4(F) and Section 106 Statements), Massachusetts Department of Public Works, Boston, MA, 1974.

R. L. Meehan, *Getting Sued and Other Tales of the Engineering Life*, The MIT Press, Cambridge, MA, 1981.

J. R. Meredith and S. J. Mantel, Jr., *Project Management: A Managerial Approach*, John Wiley & Sons, Inc., New York, NY, 1995.

J. Morgenstern, "The Fifty-nine-story Crisis," *The New Yorker*, 29 May 1995.

T. T. Nagle, *The Strategy and Tactics of Pricing*, Prentice-Hall, Englewood Cliffs, NJ, 1987.

A. Newell and H. A. Simon, *Human Problem Solving*, Prentice-Hall, Englewood Cliffs, NJ, 1972.

K. N. Otto, Measurement Methods for Product Evaluation, *Research in Engineering Design*, 7:86–101, 1995.

G. D. Oberlander, *Project Management for Engineering and Construction*, McGraw-Hill, New York, NY, 1993.

G. Pahl and W. Beitz, *Engineering Design: A Systematic Approach*, 2nd Edition, Springer, London, England, 1996.

A. Palladio, *The Four Books of Architecture*, Dover, New York, NY, 1965.

Y. C. Pao, *Elements of Computer-Aided Design and Manufacturing*, John Wiley & Sons, Inc., New York, NY, 1984.

P. Y. Papalambros and D. J. Wilde, *Principles of Optimal Design: Modeling and Computation*, Cambridge University Press, Cambridge, England, 1988.

T. E. Pearsall, *The Elements of Technical Writing*, Allyn & Bacon, Needham Heights, MA, 2001.

H. Petroski, *Design Paradigms*, Cambridge University Press, New York, NY, 1994.

H. Petroski, *Engineers of Dreams*, Alfred A. Knopf, New York, NY, 1995.

H. Petroski, *To Engineer is Human*, St. Martin's Press, New York, NY, 1985.

W. S. Pfeiffer, *Pocket Guide to Technical Writing*, Prentice Hall, Upper Saddle River, NJ, 2001.

L. Phips, *The Economics of Price Discrimination*, Cambridge University Press, Cambridge, England, 1985.

S. Pugh, *Total Design: Integrated Methods for Successful Product Engineering*, Addison-Wesley, Wokingham, England, 1991.

H. E. Riggs, *Financial and Cost Analysis for Engineering and Technology Management*, John Wiley & Sons, Inc., New York, NY, 1994.

J. L. Riggs and T. M. West, *Essentials of Engineering Economics*, McGraw-Hill, New York, NY, 1986.

E. S. Rubin, *Introduction to Engineering and the Environment*, McGraw-Hill, New York, 2001.

M. D. Rychener (Editor), *Expert Systems for Engineering Design*, Academic Press, Boston, MA, 1988.

D. G. Saari, *Basic Geometry of Voting*, Springer-Verlag, New York, 1995.

D. G. Saari, "Bad Decisions: Experimental Error or Faulty Decision Procedures," unpublished manuscript, courtesy of the author, 2001a.

D. G. Saari, *Decisions and Elections: Explaining the Unexpected*, Cambridge University Press, New York, 2001b.

M. Salvadori, *Why Buildings Stand Up*, McGraw-Hill, New York, NY, 1980.

Y. Saravanos, J. Schauer and C. Wassman, *Sliding Fulcrum Stabilizer*, E4 Project Report, Department of Engineering, Harvey Mudd College, Claremont, CA, 2000.

D. A. Schon, *The Reflective Practitioner*, Basic Books, New York, NY, 1983.

D. Schroeder, "Little Land Bruisers," *Car and Driver*, 96–109, May 1998.

R. G. Schroeder, *Operations Management: Decision Making in the Operations Function*, McGraw-Hill, New York, NY, 1993

M. J. Scott and E. K.Antonsson, "Arrow's theorem and Engineering Decision Making," *Research in Engineering Design*, *11*, 218–228, 1999.

J. J. Shah, "Experimental Investigation of Progressive Idea Generation Techniques in Engineering Design," *Proceedings of the 1998 ASME Design Theory and Methodology Conference*, American Society of Mechanical Engineers, New York, NY, 1998.

H. A. Simon, "Style in Design," in C. M. Eastman (Editor), *Spatial Synthesis in Computer-Aided Building Design*, Applied Science Publishers, London, England, 1975.

H. A. Simon, *The Sciences of the Artificial*, 3rd Edition, MIT Press, Cambridge, MA, 1996.

L. Stauffer, *An Empirical Study on the Process of Mechanical Design*, Thesis, Department of Mechanical Engineering, Oregon State University, Corvallis, OR, 1987.

L. Stauffer, D. G. Ullman and T. G. Dietterich, "Protocol Analysis of Mechanical Engineering Design," In *Proceedings of the 1987 International Conference on Engineering Design*. Boston, MA, 1987.

S. Stevenson and S. Whitmore, *Strategies for Engineering Communication*, John Wiley & Sons, Inc., New York, 2002.

G. Stevens, *The Reasoning Architect: Mathematics and Science in Design*, McGraw-Hill, New York, NY, 1990.

G. Stiny and J. Gips, *Algorithmic Aesthetics*, University of California Press, Berkeley, CA, 1978.

N. P. Suh, *Axiomatic Design: Advances and Applications*, Oxford University Press, Oxford, England, 2001.

N. P. Suh, *The Principles of Design*, Oxford University Press, Oxford, England, 1990.

M. C. Thomsett, *The Little Black Book of Project Management*, American Management Association, New York, NY, 1990.

C. Tong and D. Sriram (Editors), *Artificial Intelligence in Engineering Design, Volume I: Design Representation and Models of Routine Design*, Academic Press, Boston, MA, 1992a.

C. Tong and D. Sriram (Editors), *Artificial Intelligence in Engineering Design, Volume II: Models of Innovative Design, Reasoning about Physical Systems, and Reasoning about Geometry*, Academic Press, Boston, MA, 1992b.

C. Tong and D. Sriram (Editors), *Artificial Intelligence in Engineering Design, Volume III: Knowledge Acquisition, Commercial Applications and Integrated Environments*, Academic Press, Boston, MA, 1992c.

B. W. Tuckman, "Developmental Sequences in Small Groups," *Psychological Bulletin*, *63*, 384–399, 1965.

E. R. Tufte, *The Visual Display of Quantitative Information*, Graphics Press, Cheshire, CT, 2001.

K. Turabian, *A Manual for Writers of Term Papers, Theses, and Dissertations*, University of Chicago Press, Chicago, 1996.

D. G. Ullman, "A Taxonomy for Mechanical Design," *Research in Engineering Design*, *3*, 1992.

D. G. Ullman, *The Mechanical Design Process*, 2nd Edition, McGraw-Hill, New York, NY, 1997.

D. G. Ullman and T. G. Dietterich, "Toward Expert CAD," *Computers in Mechanical Engineering*, *6* (3), 1987.

D. G. Ullman, T. G. Dietterich and L. Stauffer, "A Model of the Mechanical Design Process Based on Empirical Data," *Artificial Intelligence for Engineering Design, Analysis and Manufacturing*, 2 (1), 1988.

D. G. Ullman, S. Wood and D. Craig, "The Importance of Drawing in the Mechanical Design Process," *Computers and Graphics*, *14* (2), 1990.

K. T. Ulrich and S. D. Eppinger, *Product Design and Development*, McGraw-Hill, New York, NY, 1995.

G. N. Vanderplaats, *Numerical Optimization Techniques for Engineering Design*, McGraw-Hill, New York, NY, 1984.

VDI, *VDI–2221: Systematic Approach to the Design of Technical Systems and Products*, Verein Deutscher Ingenieure, VDI-Verlag, Translation of the German Edition 11/1986, 1987.

C. E. Wales, R. A. Stager and T. R. Long, *Guided Engineering Design: Project Book*, West Publishing Company, St. Paul, MN, 1974.

J. Walton, *Engineering Design: From Art to Practice*, West Publishing, St. Paul, MN, 1991.

D. J. Wilde, *Globally Optimal Design*, John Wiley & Sons, Inc., New York, NY, 1978.

T. T. Woodson, *Introduction to Engineering Design*, McGraw-Hill, New York, NY, 1966.

R. N. Wright, S. J. Fenves and J. R. Harris, *Modeling of Standards: Technical Aids for Their Formulation, Expression and Use*, National Bureau of Standards, Washington, D.C., March 1980.

C. Zener, *Engineering Design by Geometric Programming*, Wiley-Interscience, New York, NY, 1971.

C. Zozaya-Gorostiza, C. Hendrickson and D. R. Rehak, *Knowledge-Based Process Planning for Construction and Manufacturing*, Academic Press, Boston, MA, 1989.

INDEX